The Business of WiMAX

The Business of WiMAX

Deepak Pareek
Resource4Business, India

JOHN WILEY & SONS, LTD

Other Wiley Editorial Offices

John Wiley & Sons Inc., 111 River Street, Hoboken, NJ 07030, USA

Jossey-Bass, 989 Market Street, San Francisco, CA 94103-1741, USA

Wiley-VCH Verlag GmbH, Boschstr. 12, D-69469 Weinheim, Germany

John Wiley & Sons Australia Ltd, 42 McDougall Street, Milton, Queensland 4064, Australia

John Wiley & Sons (Asia) Pte Ltd, 2 Clementi Loop #02-01, Jin Xing Distripark, Singapore 129809

John Wiley & Sons Canada Ltd, 22 Worcester Road, Etobicoke, Ontario, Canada M9W 1L1

Wiley also publishes its books in a variety of electronic formats. Some content that appears in print
may not be available in electronic books.

British Library Cataloguing in Publication Data
A catalogue record for this book is available from the British Library

ISBN-13 978-0-470-02691-5 (HB)
ISBN-10 0-470-02691-X (HB)

Typeset in 10.5/13pt Sabon-Roman by Thomson Press (India) Limited, New Delhi, India
Printed and bound in Great Britain by TJ International Ltd, Padstow, Cornwall
This book is printed on acid-free paper responsibly manufactured from sustainable forestry in which
at least two trees are planted for each one used for paper production.

Contents

Preface

Modern technologies are revolutionizing the way we work, play and interact. It is not an exaggeration to suggest that these disruptive technologies are altering the way we live and that, with every passing day, these disruptions are becoming greater. This trend has created new competitive threats as well as new opportunities in every walk of life.

The telecommunications industry is finding itself most affected by these developments. The human behavioural trait responsible for these unprecedented but welcoming tremors is 'communication'.

WiMAX: THE SUBJECT

When this book was initially planned, WiMAX technology was new, unproven and the subject of much doubt and uncertainty. It held promise to be sure, but then so have many broadband wireless formats over the years, with little result. WiMAX is still very new and unproven, having only just seen initial deployments (in fixed formats) in service provider networks. The big difference, however, and the reason why WiMAX remains an area of technological promise is the push for standardization and equipment certification that has occurred over recent years. Even though commercial deployment and service introduction are still in their infancy, WiMAX is in a standardization and certification cycle that will potentially lower the costs of equipment production and pave the way for future iterations of the technology to be released and deployed more efficiently.

To truly understand WiMAX, one must wade through a mass of technical talk, market speak and plain old hype about what the specification can and cannot do. This book will outline how 802.16 technology works, how it is likely to evolve, what is driving the technology, what it will realistically deliver and, just as critically, what it cannot achieve. It will also look at the equipment necessary to bring the market to life and analyse potential critics of the technology such as incumbent wireline and wireless network operators, competitive DSL suppliers and enterprises. The likely time frames for commercial availability of WiMAX products will be examined and the obstacles that stand in the way of widespread WiMAX adoption, ranging from spectrum issues to the competitive threat of 802.11 and 3G cellular technologies. This book will also look at how pricing pressures will be intense in this market, even for first-generation products.

The book will draw on some of the original research reports from a leading WiMAX analysis and research outfit, Research4Business (R4B), which provide the most comprehensive and up-to-date analysis ever undertaken on the WiMAX market. Based on information from direct interviews conducted with technology suppliers, service providers and investors with a direct interest in the WiMAX market, these reports provide an overview of everything linked to WiMAX. Having a grasp on this information was vital for as an author on this subject, and I am grateful to R4B.

The Business of WiMAX is a book which is a step in the direction of demystifying WiMAX. It is divided into four sections, each covering an important aspect of the subject. The heart of the book is an in-depth exploration of the business case for WiMAX, WiMAX business models and success strategies for the players.

Part One – Understanding WiMAX

This section provides an overview of WiMAX landscape and consists of three chapters.

Part Two – WiMAX Effect

This section provides a top-level view of solutions provided by WiMAX, their applications and the impact and consists of three chapters.

Part Three – WiMAX Business

This section gives a business perspective of the subject, which includes a top-level view of the market, opportunity and economics of the subject and consists of three chapters.

Part Four – WiMAX Strategy

This section discusses Strategies for Success for various stakeholders and players in the WiMAX economy; it consists of four chapters, each covering a specific player.

Some detailed readings are also included on different topics that have been touched upon in the main text but not covered in full so as not to limit the audience. The Appendix provides a behind-the-scenes look at the process for certification, from technological tests right down to how the gear will be labelled once it is certified, as well as present proprietary broadband technologies and key market players.

Get Ready
The WiMAX is coming...

PART One
Understanding WiMAX

1

Introduction

The start of the new millennium is witnessing a telecommunications world that is very different from even the recent past. The huge explosion of wireless and broadband technologies over the last few years has captured the imagination and innovativeness of technologists around the world.

It has been a constant human endeavour to communicate more effectively and at the same time to be free of any bondage, physical or psychological. A similar underlying trend can also be seen in the evolution of telecommunications. The need for mobility and higher speeds in an ever-changing environment has been of paramount importance.

Clearly with ever-increasing expectation and highly dynamic technologies, challenges lie ahead, driven by the intrinsic human trait of nomadicity and the fundamental need to communicate in a feature-rich environment. With the new-found power of mobility and broadband, the telecommunications industry has tapped into an explosive technology mix that can grow exponentially once creativity and innovativeness come into play (Figure 1.1).

1.1 WIRELESS COMMUNICATION: ANY TIME, ANY PLACE

The main factor behind this tremendous growth has been the wireless medium's ability to satisfy substantially any two components of the three that comprise the ultimate goal of telecommunications: any

The Business of WiMAX Deepak Pareek
© 2006 John Wiley & Sons, Ltd

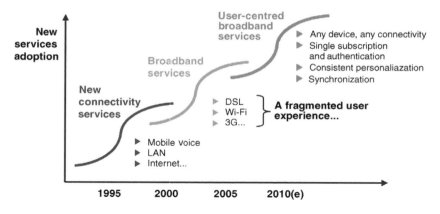

Figure 1.1   Connectivity evolution

information, any time, any place. Wireless communications systems provide anytime, anywhere communications.

The future of wireless lies in faster, more reliable methods of transferring data and, to a lesser extent, increased use of voice commands and audio improvements.

Some of the inherent characteristics of wireless communications systems which make it attractive for users, are discussed below in detail.

Mobility

Wireless systems enable better communication, enhanced productivity and better customer service. A wireless communications system allows users to access information beyond their desk and conduct business from anywhere.

Reach

Wireless communications systems enable people to be better connected and reachable wherever they are.

Simplicity

Wireless communications systems are faster and easier to deploy than cabled networks. Installation can take place simply, ensuring minimum disruption.

Flexibility

Wireless communications systems provide flexibility, as a subscriber can have full control of his/her communication.

Setup Cost

The initial costs of implementing a wireless communications system compare favourably with those for a traditional wireline or cable system. Communications can reach areas where wiring is infeasible or costly, e.g. rural areas, old buildings, battlefields, vehicles.

Falling Services Cost

Wireless service pricing is rapidly approaching wireline service pricing.

Global Accessibility

Roaming makes the dream of global accessibility a reality, since today most parts of the globe are well covered by a wireless service provider. Roaming services also allow the flexibility to stay connected anywhere.

Smart

Wireless communications systems provide new smart services like SMS and MMS.

Cultural

Wireless communications systems comprise personal devices, whereas wireline is more connected to a place, e.g. the office. In today's world wireless communication is no longer just about cell phones; instead it is the direction in which telecommunications seem to be heading to provide all possible ways of keeping information place-independent to a greater or lesser extent (Figure 1.2).

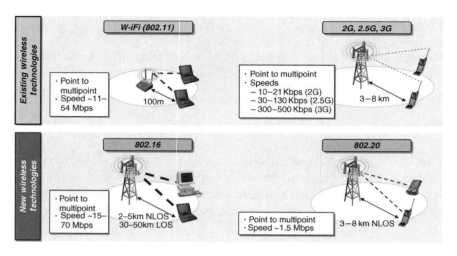

Figure 1.2   Wireless technologies

1.2   WIRELESS NETWORKS

The tetherless nature of connectivity provides its users with almost unrestricted mobility and the facility to access the network from anywhere. While in a wired network an address represents a physical location; in a wireless network the addressable unit is a station, which is the destination for a message and is not (necessarily) at a fixed location. Although wireless networks have been around for sometime, they are gaining popularity rapidly with standardization and reductions in the cost of hardware components.

What Does 'Wireless' Mean?

The world is going wireless, with an ever-increasing number of people reaping the rewards of wireless communications. From mobiles to laptops and personal digital assistants (PDAs), the list of wireless technological devices is endless.

'Wireless' means transmitting signals using radio waves as the medium instead of wires. Remote controls for television and other customer electronic appliances were the first wireless devices to become part of everyday life. Now cordless keyboards and mice, PDAs, and digital and cellular phones are commonplace.

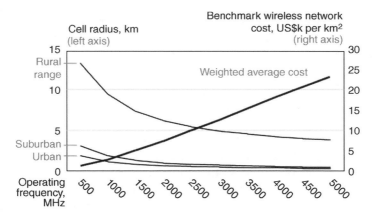

Figure 1.3   Wireless network: coverage vs cost

Wireless technologies are used for tasks as simple as switching off the television or as complex as supplying the sales force with information from an automated enterprise application while in the field. For businesses, wireless technologies mean new ways to stay in touch with customers, suppliers and employees (Figure 1.3).

Wireless Economics

The most notable factors that have contributed to this exponential growth are the Internet boom, the need for mobility in an ever-changing environment, low costs (flat rate), increased data rates, increased battery life, application friendliness and innovativeness. In many countries there are now more wireless phone, lines than fixed lines. There are a number of reasons for this unexpected boom in wireless networks, the foremost being the use of wireless or mobile phones, which is more convenient and requires less investment than a fixed infrastructure.

In addition, a wireless infrastructure has more 'flexibility' than a fixed infrastructure, in which at least the part of the access network closest to the user is dedicated to a specific locations and its profitability depends on the use made of this access by that household or business. Wireless networks do not suffer from this limitation; their use can be shared and reassigned much more easily, and they can become profitable more rapidly.

Some analysts of the telecommunications industry believe that, within a few years, most telephone calls in the residential market will be placed over wireless networks.

Drivers for wireless networks

It is by improving business processes that wireless access will find a place in many enterprises. Several internal and external factors are converging to drive a sense of urgency among businesses to find these process efficiencies, for example increased customer expectations, need for effective time utilization and employee empowerment, cost reduction and cost avoidance, advancing enterprise connectivity, legislation and government requirements.

Issues for wireless networks

As with any relatively new technology, there are many issues that affect the implementation and utilization of wireless networks. These are both common and specific, depending on the type of wireless network. Some of the common factors include electromagnetic interference and physical obstacles that limit coverage of wireless networks, while others are more specific, such as standards, data security, throughput and ease of use, (Figure 1.4).

Network Topology

There are basically three ways to connect a wireless network.

Point-to-point bridge

A bridge connects two networks. A point-to-point bridge would interconnect two buildings. Access points connect a network to multiple users. For example, a wireless LAN bridge can interface with an Ethernet network directly to a particular access point. This may be necessary if you have several devices in a distant part of the facility that are interconnected using Ethernet.

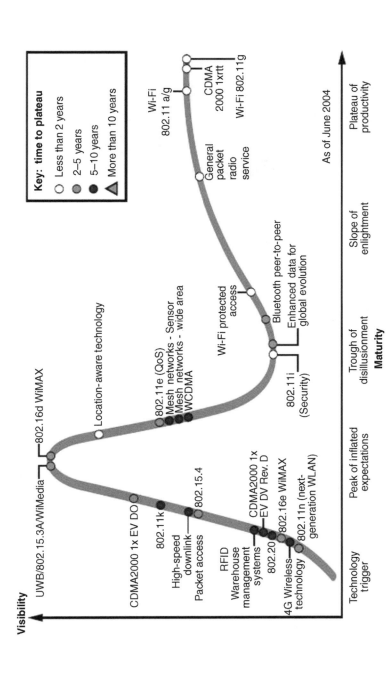

Figure 1.4 Wireless hype cycle. Reproduced from P. Redmant *et al.*, *Hype Cycle for Wireless Networking, 2004* by permission of Gartner Inc.

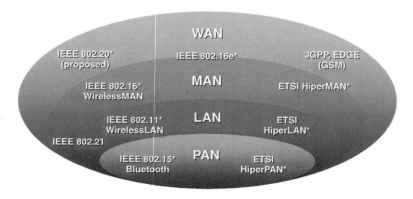

Figure 1.5   Wireless network standards

Point-to-multipoint bridge

When connecting three or more LANs that may be located on different floors in a building or across buildings, the point-to-multipoint wireless bridge is utilized. The multipoint wireless bridge configuration is similar to a point-to-point bridge in many ways.

Mesh or *ad hoc* network

An *ad hoc* (peer-to-peer) network is an independent local area network that is not connected to a wired infrastructure and in which all stations are connected directly to one another (called a mesh topology). Configuration of a WLAN in *ad hoc* mode is used to establish a network where wireless infrastructure does not exist or where services are not required, such as a trade show or collaboration by co-workers at a remote location (Figure 1.5).

1.3   WIRELESS TECHNOLOGIES

Wireless networking technologies range from global voice and data networks to infrared light and radio frequency technologies optimized for short-range wireless connections. Devices commonly used for wireless networking include portable computers, desktop computers, handheld computers, PDAs, cellular phones, pen-based computers and pagers. Wireless technologies have evolved substantially over the past few years and, depending on their range, can be classified in different ways.

Wireless Wide Area Network

This network is designed to enable users to access the Internet via a wireless wide area network (WWAN) access card and a PDA or laptop. Data speeds are very fast compared with the data rates of mobile telecommunications technology, and their range is also extensive. Cellular and mobile networks based on CDMA and GSM are good examples of WWAN.

Wireless Local Area Network

This network is designed to enable users to access the Internet in localized hotspots via a wireless local area network (WLAN) access card and a PDA or laptop. While data speeds are relatively fast compared with the data rates of mobile telecommunication technology, their range is limited. Among the various WLAN solutions, Wi-Fi is the most widespread and popular.

Wireless Personal Area Network

This network is designed to enable the users to access the Internet via a wireless personal area network (WPAN) access card and a PDA or laptop. While data speeds are very fast compared with the data rates of mobile telecommunications technology, their range is very limited.

Wireless Region Area Network

This network is designed to enable the users to access the Internet and multimedia streaming services via a wireless region area network (WRAN). Data speeds are very fast compared with the data rates of mobile telecommunication technology as well as other wireless network, and their range is also extensive. The specific charter of the WRAN working group is to 'develop a standard for a cognitive radio-based air interface for use by license-exempt devices on a non-interfering basis in spectrum that is allocated to the TV Broadcast Service'. WRAN, which is presently in its infant stage, is the most recent addition to a growing list of wireless access network acronyms defined by coverage area.

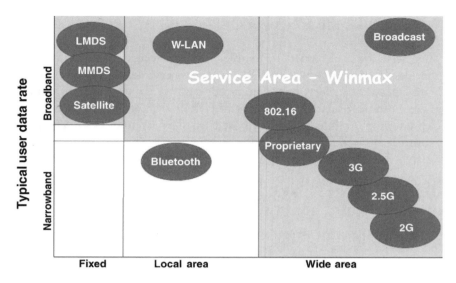

Figure 1.6   Wireless technologies, bandwidth versus distance

We will deal with two types of wireless networks, WLAN and WWAN (Figure 1.6).

1.4   THIRD-GENERATION MOBILE SYSTEMS

The remarkable growth of cellular mobile telephony as well as the need for wireless data services promises an impressive potential for a market that combines high speed with the convenience of mobile technology. Universal Mobile Telecoms Service (UMTS) Generation 3 will offer users an alternative with high-speed access, allowing connectivity from any location on the planet (Table 1.1).

First-generation Mobile Systems

The first wireless generation introduced analogue systems transmitting over radio frequencies, used primarily for voice. The operation of first-generation mobile phones was based on analogue radio technology. It was composed of three elements – mobile telephone, cell sites and mobile switching centres (MSC). The system was designed using two different radio channels. The first was the control channel, and the second was the voice channel.

Table 1.1 Various cellular and mobile technologies

Technology	Maximum speed	Launch	Benefits	Drawbacks	Bottom-line
GPRS	171.2 kbps	2001	Packet data for GSM world	Data rates may disappoint	Will be the most successful technology through 2005
HSCSD	115 kbps	In use	Dedicated channels	Low deployment, expensive	Will not be mainstream
EDGE Classic	384 kbps	2003	Higher data rates for both packet and circuit	Expensive, little terminal support	Will not be able to compete with W-CDMA
EDGE Compact	250 kbps	2002	Higher data rates for both packet and circuit TDMA networks	AT&T (main proponent) has changed direction	Unlikely to be successful
CDMA/ IS-95B	115 kbps	In use	Interim packet technology for CDMA networks, backward compatible with IS-95A	Only adopted in Japan and South Korea	Most carriers will prefer to deploy Cdma2000 1×MC
Cdma2000 1×MC	307 kbps	2002	High data rates, smooth migration path	Limited global footprint	Good technology but will not survive.
PDC-P	9.6 kbps	In use	Used by NTT DoCoMo	Japan only, low data rate	Currently the most successful wireless packet technology in the world
W-CDMA	2 Mbps	2001– 2003	Massive industry support	High licence fee	*De facto* global standard
Cdma2000 3×MC	2 Mbps	2004	Backward compatible with 1×MC and IS-95A	Support has cooled down	Good technology but unlikely to be successful
CDMA 1 EVDV	2.4 Mbps	2003	Smooth migration path	Limited global footprint	Will not be mainstream
CDMA 1 XTREME	5.2 Mbps	2004	Very high data rates	Proprietary – Motorola, Nokia	No indication of intent from carriers

The control channel was responsible for carrying digital messages, which allowed the phone to retrieve system control information and compete for access. It used frequency shift keying modulation (FSK) to complete this task. The responsibility of voice channels was to transmit voice data over an analogue signal using frequency modulation (FM) radio.

Second-generation Mobile Systems

Compared with first-generation systems, second-generation (2G) systems use digital multiple access technology, such as time division multiple access (TDMA) and code division multiple access (CDMA). The global system for mobile communications or GSM uses TDMA technology to support multiple users.

Examples of second-generation systems are GSM, cordless telephones (CT2), personal access communications systems (PACS) and digital European cordless telephones (DECT). A new design was introduced into the mobile switching centre of second-generation systems. In particular, the use of base station controllers (BSCs) lightened the load placed on the MSC found in first-generation systems. This design allows the interface between the MSC and BSC to be standardized. Hence, considerable attention was devoted to interoperability and standardization in second-generation systems so that carrier could employ different manufacturers for the MSC and BSC.

In addition to enhancements in MSC design, the mobile-assisted handoff mechanism was introduced. By sensing signals received from adjacent base stations, a mobile unit can trigger a handoff by performing explicit signalling with the network. Second-generation protocols use digital encoding and include GSM, D-AMPS (TDMA) and CDMA (IS-95). The protocols behind 2G networks support voice and some limited data communications, such as fax and short messaging service (SMS), and most 2G protocols offer different levels of encryption and security. While first-generation systems support primarily voice traffic, second-generation systems support voice, paging, data and fax services (Figure 1.7).

2.5G Mobile Systems

The move into the 2.5G world began with the idea of providing decent data connectivity without substantially changing the existing

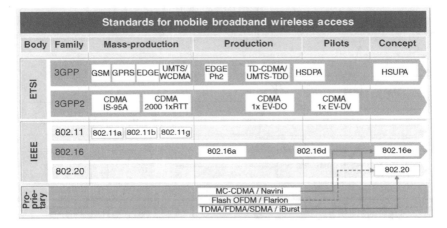

Figure 1.7   Standards – broadband wireless access

2G infrastructure. Some of the cellular technologies capable of achieving this goal are discussed below.

High-speed circuit-switched data (HSCSD)

High-speed circuit-switched data (HSCSD) were designed to allow GSM
networks transfer data at rates up to four times the original network
data rates.

General packet radio service (GPRS)

General Packet radio service (GPRS) is a radio technology for GSM
networks that provides packet-switching protocols, shorter setup time
for ISP connections, increased data rates as well as charging based on the
amount of data transferred rather than the time spent in transferring the
data.

The next generation of data, heading towards third-generation and
personal multimedia environments, is built on GPRS and is known as
enhanced data rate for GSM evolution (EDGE).

Enhanced data GSM environment (EDGE)

EDGE allows GSM operators to use existing GSM radio bands
to offer wireless multimedia IP-based services and applications at

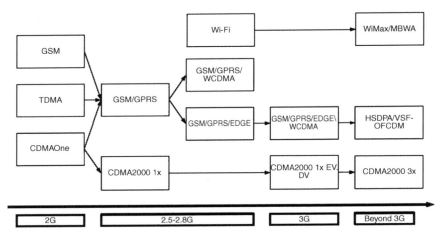

Figure 1.8 Migration path from 2G to beyond 3G

theoretical maximum speeds of 384 kbps (up to a theoretical maximum of 554 kbps) with a bit-rate of 48 kbps per time slot and up to 69.2 kbps per time slot under good radio conditions. EDGE also allows operators to function without a 3G licence and competes with 3G networks offering similar data services and, in some cases, challenging 3G data rates.

Implementing EDGE is relatively painless and requires relatively small changes to network hardware and software as it uses the same TDMA frame structure, logic channel and 200 kHz carrier bandwidth as GSM networks. Designed to coexist with GSM networks and 3G WCDMA (Figure 1.8), EDGE offers data rates equivalent to ATM-like speeds of up to 2 Mbps.

Third-generation Mobile Systems

Third-generation mobile systems are faced with several challenging technical issues, such as the provision of seamless services across both wired and wireless networks. In Europe, there are two evolving networks under investigation: UMTS (Universal Mobile Telecommunications Systems) and MBS (Mobile Broadband Systems).

CDMA2000

Emerging out of the standard IS-95, CDMA2000 has already undergone a considerable amount of development, particularly in the area of multi-

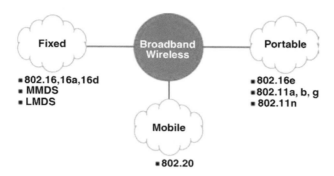

Figure 1.9   Broadband wireless access technologies

channel working. Operators of narrowband CDMA One (IS-95A/B) can deploy services designated as 3G in existing as well as new spectrum bands.

Wideband code division multiple access (W-CDMA)

Many see W-CDMA technology as the preferred platform for the 3G cellular systems, since it offers seamless migration for GSM networks (which may or may not have already progressed to general packet radio service/enhanced data for GSM evolution – GPRS/EDGE technology) and can provide a migration path for narrow-band CDMA networks. Thus, W-CDMA will be able to cover much of the world with its comprehensive backward compatibility to such networks (Figure 1.9).

1.5   WLAN

WLAN is an acronym for wireless local area network, also referred to as LAWN. It is a type of local area network that uses high-frequency radio waves rather than wires to communicate between nodes.

WLANs are slowly but surely taking hold in homes, small businesses and corporations. When you compare the cost of WLAN interface cards and access point with that of wiring up a cubicle and the inflexibility of that wired connection, it is easy to see why people are attracted to WLANs, although one has to take into consideration the backhaul which connects to the access point (Table 1.2).

Performance ranges from standard Ethernet performance down to perhaps 2 Mbps if there is significant interference or if the user strays too far from an access point. If the network interface card (NIC) and access point support roaming, a user can wander around a building or campus

Table 1.2 Various WLAN technologies

Standard	Data rate	Modulation scheme	Security	Pros/cons and more information
IEEE 802.11	Up to 2 Mbps in the 2.4 GHz band	FHSS or DSSS	WEP and WPA	This specification has been extended into 802.11b.
IEEE 802.11a (Wi-Fi)	Up to 54 Mbps in the 5 GHz band	OFDM	WEP and WPA	Products that adhere to this standard are considered 'Wi-Fi Certified'. Eight available channels. Less potential for RF interference than 802.11b and 802.11g. Better than 802.11b at supporting multimedia voice, video and large-image applications in densely populated user environments. Relatively shorter range than 802.11b. Not interoperable with 802.11b.
IEEE 802.11b (Wi-Fi)	Up to 11 Mbps in the 2.4 GHz band	DSSS with CCK	WEP and WPA	Products that adhere to this standard are considered 'Wi-Fi Certified'. Not interoperable with 802.11a. Requires fewer access points than 802.11a for coverage of large areas. Offers high-speed access to data at up to 300 feet from base station. 14 channels available in the 2.4 GHz band (only 11 of which can be used in the US due to FCC regulations) with only three non-overlapping channels.

IEEE 802.11g (Wi-Fi)	Up to 54 Mbps in the 2.4 GHz band	OFDM above 20 Mbps, DSSS with CCK below 20 Mbps	WEP and WPA	Products that adhere to this standard are considered 'Wi-Fi Certified'. May replace 802.11b. Improved security enhancements over 802.11. Compatible with 802.11b. 14 channels available in the 2.4 GHz band (only 11 of which can be used in the US due to FCC regulations) with only three non-overlapping channels.
EEE 802.16 (WiMAX)	Specifies WiMAX in the 10–66 GHz range	OFDM	DES3 and AES	Commonly referred to as WiMAX or less commonly as WirelessMAN or the Air Interface Standard, IEEE 802.16 is a speci-fication for fixed broadband wireless metropolitan access networks (MANs).
IEEE 802.16a (WiMAX)	Added support for the 2–11 GHz range	OFDM	DES3 and AES	Commonly referred to as WiMAX or less commonly as WirelessMAN or the Air Interface Standard, IEEE 802.16 is a specification for fixed broadband wireless metropolitan access networks (MANs).
Bluetooth	Up to 2 Mbps in the 2.45 GHz band	FHSS	PPTP, SSL or VPN	No native support for IP, so it does not support TCP/IP and wireless LAN applications well. Not ori-ginally created to support wireless LANs. Best suited to connecting PDAs, cell phones and PCs in short intervals.

(Continued)

Table 1.2 (*Continued*)

Standard	Data rate	Modulation scheme	Security	Pros/cons and more information
HomeRF	Up to 10 Mbps in the 2.4 GHZ band	FHSS	Independent network IP addresses for each network. Data are sent with a 56-bit encryption algorithm.	**Note:** HomeRF is no longer being supported by any vendors or working groups. Intended for use in homes, not enterprises. Range is only 150 feet from base station. Relatively inexpensive to set up and maintain. Voice quality is always good because it continuously reserves a chunk of bandwidth for voice services. Responds well to interference because of frequency-hopping modulation.
HiperLAN/1 (Europe)	Up to 20 Mbps in the 5 GHz band	CSMA/CA	Per session encryption and individual authentication.	Only in Europe. HiperLAN is totally *ad hoc*, requiring no configuration and no central controller. Does not provide real isochronous services. Relatively expensive to operate and maintain. No guarantee of bandwidth.
HiperLAN/2 (Europe)	Up to 54 Mbps in the 5 GHz band	OFDM	Strong security features with support for individual authentication and per-session encryption keys.	Only in Europe. Designed to carry ATM cells, IP packets, fire wire packets (IEEE 1394) and digital voice (from cellular phones). Better quality of service than HiperLAN/1 and guarantees bandwidth.
OpenAir	Pre-802.11 protocol, using frequency hopping and 0.8 and 1.6 Mbps bit rate	CSMA/CA with MAC retransmissions	OpenAir does not implement any encryption at the MAC layer, but generates Network ID based on a password (Security ID).	OpenAir is the proprietary protocol from Proxim. All OpenAir products are based on Proxim's module.

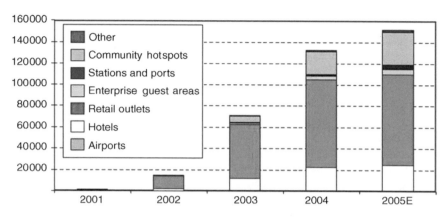

Figure 1.10 Wi-Fi hotspots by type of location

and the NIC will automatically switch between access points based on the strength of the beacon signal it receives from nearby access points. The strongest signal wins.

Growth in WLANs can be traced to the creation of 802.11, the IEEE technical standard that enabled high-speed mobile interconnectivity. After sustained efforts by the WLAN Standards Working Group, the IEEE ratified a new rate standard for WLANs, viz. 802.11b, also known as wireless fidelity (Wi-Fi). Some other WLAN technologies are WiMAX, Bluetooth, HomeRF and Open Air (Figure 1.10).

What is Wi-Fi?

Wi-Fi was used to refer only to the 802.11b standard, but now also refers to the broader spectrum of WLAN standards, including 802.11a and the emerging 802.11g. This standard was certified by the Wireless Ethernet Compatibility Alliance (WECA). The 802.11a standard – approved by the IEEE at the same time as 802.11b – provides for data rates up to 54 Mbps at 5 GHz frequency. The 802.11g standard, with an even higher data rate, has recently been introduced and operates at the same frequency as 802.11b. Of all these emerging standards, 802.11b has been the most widely deployed.

The 802.11b standard works at the 2.4 GHz frequency of the electromagnetic spectrum and allows users to transmit data at speeds up to 11 Mbps. However, a number of wireless products, such as cordless phones and garage door openers, also use the 2.4 GHz frequency and can cause disruptions in the service.

The 802.11a standard works on the 5 GHz frequency, which is relatively uncluttered and allows data transfer rates up to 54 Mbps, but has a shorter effective range than 802.11b at about 15–22.5 m.

Hotspots

A major trend found with WLANs is the establishment of WLAN 'hotspots' and community networks. WLAN products are now being shipped that have performances of 54 Mbps or more.

A hotspot provides a WLAN service, for free or for a fee, from a wide variety of public meeting areas, including coffee shops and airport lounges. There are currently thousands of hotspots worldwide and new access points are being added daily. To use hotspots, your laptop must be configured with Wi-Fi Certified^{TM} technology so that you can connect with other products. Wi-Fi-CERTIFIED laptops can send and receive data anywhere within the range of a WLAN base station.

Currently, the Institute of Electrical and Electronics Engineers (IEEE) is preparing the final specification for 802.11g, which combines the use of the 2.4 GHz frequency with the faster download speeds offered by 802.11a. Products are already available based on the draft standard, and any changes made during the final process between now and the middle of 2005 will require just a software update, according to vendors and the Wi-Fi Alliance.

Problems with Wi-Fi

The sluggish adoption of Wi-Fi technology in its initial stages was not due to any specific issue; the Wi-Fi bandwagon was derailed because of multiple problems. Foremost among them was the lack of security, which meant that wireless networks can make sensitive corporate information available to outsiders. In addition, several different standards, versions and products were causing great confusion, and not all products worked with all standards. Let us examine each of these issues in detail.

Security Security concerns have held back Wi-Fi adoption in the corporate world. Hackers and security consultants have demonstrated how easy it can be to crack the current security technology, known as wired equivalent privacy (WEP), used in most Wi-Fi connections. A hacker can break into a Wi-Fi network using readily available materials and software.

The IEEE is currently working towards the release of 802.11i, which is a software standard that seeks to improve security features in various 802.11 wireless hardware standards. The purpose of 802.11i is to improve the safety of transmissions (management and distribution of the keys, coding and authentification). This standard rests on the Advanced Encryption Standard (AES) and proposes coding communications for transmission using technologies 802.11a, 802.11b and 802.11g.

As a stopgap measure for Wi-Fi users until a new software standard from the IEEE is ratified, a new security technology known as Wi-Fi protected access (WPA) has been commissioned. In an attempt to allay security concerns, the Wi-Fi alliance has taken up the initiative to certify Wi-Fi products for WPA. Products certified for WPA will feature several technologies not found in WEP, including improved key management technology and temporal key integrity protocol (TKIP). Users of current Wi-Fi products will be able to upgrade to WPA through software updates.

When the final version of 802.11i is ratified by the IEEE, it will contain a security protocol known as counter with cipher block chaining message authentication code protocol (CCMP). This will add an additional layer of security for the second version of WPA based on the completed standard.

Compatibility and Interpretability   One of the bigger problems with Wi-Fi is compatibility and interpretability, for example 802.11a products are not compatible with 802.11b products, due to the different operating frequencies, and 802.11a hotspots would not help a 802.11b client. Also, owing to lack of standardization, harmonization and certification, different vendors come out with products that do not work with each other. These problems have plagued Wi-Fi growth for a long time, and the Wi-Fi alliance is looking for solutions.

Billing Issues   Wi-Fi vendors are also looking for ways to solve the problem of back-end integration and billing that has dogged the roll-out of commercial Wi-Fi hotspots. One model that can be used as the starting point is the way cell phone carriers have set up their back-end billing systems. However, the amount of capital required to set up a Wi-Fi hotspot is far less than that required for cellular operators (about $100 for a wireless base station vs about $1 million for a cell phone tower); hence, the economic scale will be quite different.

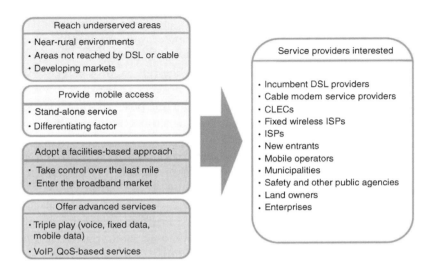

Figure 1.11   What can WiMAX deliver

Some of the ideas under consideration for Wi-Fi billing include per day, per hour and unlimited monthly connection fees. Right now, users are willing to pay a premium for hotspot access, but as pricing becomes more competitive, hotspot owners will need a larger share of the revenues they generate for the equipment companies and hotspot providers.

Right now the most prevalent revenue share models value the network far more than the location. The owner of a venue hosting a hotspot receives substantially lower percentage of the revenue generated by the Wi-Fi hotspot, while the major part goes to the equipment manufacturer and the hotspot provider, which is responsible for support and installation.

Larger venues such as airports or convention centres can make a handsome amount even with a small share of the revenue, but venues like coffee shops, pubs, malls and fast-food outlets are not making attractive amounts from Wi-Fi hotspots. These venues being the key to driving Wi-Fi growth as well as penetration, hotspot providers and aggregators will need to offer them a larges slice of the pie to encourage more venues to install hotspots (Figure 1.11).

1.6   WiMAX

The prospect of broadband Internet access anywhere, at any time, has seemed a distant dream, far from reality for the vast majority of PC, laptop and handheld users. However, with WiMAX, it will soon become something users cannot live without. WiMAX is one of the hottest wireless technologies around today.

WiMAX systems are expected to deliver broadband access services to residential and enterprise customers in an economical way. Although it has one name, WiMAX will be two different market technologies. The first is for fixed wireless and falls under the IEEE 802.16-2004 standard approved last year. The second, for mobile applications, will be under the 802.16e specification expected to be finalized this year.

As of now, fixed WiMAX is capable of becoming a replacement for DSL or cable or for network backhaul. In future, WiMAX will transform the world of mobile broadband by enabling the cost-effective deployment of metropolitan area networks based on the IEEE 802.16e standard to support notebook PC and mobile users on move.

There are many advantages of systems based on 802.16, e.g. the ability to provide service even in areas that are difficult for wired infrastructure to reach and the ability to overcome the physical limitations of traditional wired infrastructure. The standard will offer wireless connectivity of up to 30 miles. The major capabilities of the standard are its widespread reach, which can be used to set up a metropolitan area network, and its data capacity of 75 Mbps.

This high-speed wireless broadband technology promises to open new, economically viable market opportunities for operators, wireless Internet service providers and equipment manufacturers. The flexibility of wireless technology, combined with high throughput, scalability and long-range features of the IEEE 802.16 standard helps to fill the broadband coverage gaps and reach millions of new residential and business customers worldwide (Figure 1.12).

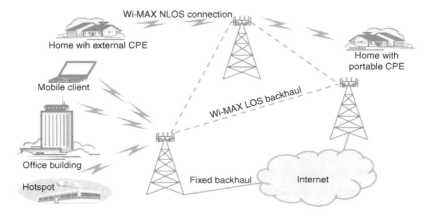

Figure 1.12   WiMAX solutions

What is WiMAX?

WiMAX is:

- A wireless technology optimized for the delivery of IP centric services over a wide area.
- A certification that denotes interoperability of equipment built to the IEEE 802.16 or compatible standard. The IEEE 802.16 Working Group develops standards that address two types of usage models: a fixed usage model (IEEE 802.16-2004) and a portable usage model (802.16 REV E, scheduled for ratification in 2005).
- A scaleable wireless platform for constructing alternative and complementary broadband networks.

Commonly referred to as WiMAX or less commonly as Wireless-MAN^{TM} or the Air Interface Standard, IEEE 802.16 is a specification for fixed broadband wireless metropolitan access networks (MANs) that use a point-to-multipoint architecture. Published on 8 April 2002, the standard defines the use of bandwidth between the licensed 10 and 66 GHz and between the 2 and 11 GHz (licensed and unlicensed) frequency ranges, and defines a MAC layer that supports multiple physical layer specifications customized for the frequency band of use and their associated regulations. 802.16 supports very high bit rates in both uploading to and downloading from a base station up to a distance of 30 miles, in order to handle such services as VoIP, IP connectivity and TDM voice and data.

Loosely, WiMAX is a standardized wireless version of Ethernet intended primarily as an alternative to wire technologies (such as cable modems, DSL and T1/E1 links) to provide broadband access to customer premises. This application is often called wireless last/first-mile broadband because the transmission distances involved are typically of this order, and the engineering problem is to bridge the final gap between the customer premises and the telco's or service provider's main network. The technology is specified by the IEEE, as the IEEE 802.16 standard.

More strictly, WiMAX is the Worldwide Microwave Interoperability Forum, a non-profit industrial body dedicated to promoting the adoption of this technology and ensuring that different vendors' products will interoperate. WiMAX will do this through developing conformance and interoperability test plans, selecting certification laboratories and hosting interoperability events for 802.16 equipment vendors. WiMAX is such a convenient term that people tend to use it for the 802.16 standards and technology themselves, although strictly it applies only to systems that meet specific conformance criteria laid down by the WiMAX Forum.

The 802.16 standard is large, complicated and evolving, and offers many options and extensions, so interoperability is a major issue that must be addressed. In particular, one extension known as 802.16a became the focus of much industry attention because it is the easiest and most useful to implement. It is likely that when people talk loosely of WiMAX they are referring to the technology for fixed wireless specified by 802.16a and its later version 802.16d.

802.16 is one of a family of technologies being standardized by the IEEE [with other bodies, such as the European Telecommunications Standards Institute (ETSI), whose Hiperman standard is harmonized with 802.16] to create wireless versions of Ethernet that can operate over distances from a few metres to tens of kilometres – from personal area networks (PANs), through LANs and MANs, to wide area networks (WANs). 802.16 is the MANs member of the family (Figure 1.13).

WiMAX Craze: Is it Misplaced?

What excites the industry is the combination of potential low cost and flexibility that WiMAX promises. In principle, WiMAX broadband

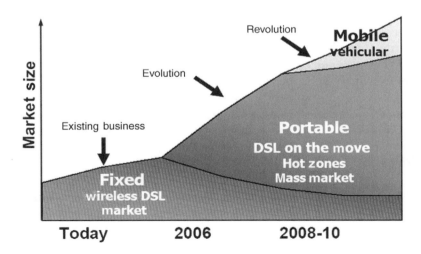

Figure 1.13 WiMAX, from evolution to revolution

networks can be built quickly and (compared with wireline systems) at relatively low cost by installing just a few wireless base stations mounted on buildings or poles to provide coverage to the surrounding area. The use of wireless eliminates the costly trenching and cabling of new wire/fibre networks, and Ethernet itself has a long history of achieving lower equipment costs than competing technologies. WiMAX networks should scale well, as extra channels and base stations can be added incrementally as bandwidth demand grows.

WiMAX is expected to offer initially up to about 40 Mbps capacity per wireless channel for both fixed and portable applications, depending on the particular technical configuration chosen, enough to support hundreds of businesses with T-1 speed connectivity and thousands of residences with DSL speed connectivity. WiMAX can support voice and video as well as Internet data.

Its first application will be to provide wireless broadband access to buildings, either in competition to existing wired networks or alone in currently unserved rural or thinly populated areas. It can also be used to connect WLAN hotspots to the Internet. WiMAX is also intended to provide broadband connectivity to mobile devices. It would not be as fast as in these fixed applications, but expectations are for about 15 Mbps capacity in a 3 km cell coverage area.

Semiconductor vendors like Intel envisage WiMAX-enabled chips appearing in PCs and other end-user equipment by 2006 and in PDAs and mobile phones by 2007 or 2008, raising the prospect of a mass market of users developing as it has for WLAN local Ethernet.

With WiMAX users could really cut free from today's Internet access arrangements and be able to go online at broadband speeds, almost wherever they like from within a MetroZone.

Millions of new subscribers worldwide will benefit from broadband access services delivered over wireless networks (Figure 1.14).

WiMAX, Wi-Fi and 3G: Friends or Foes

WiMAX wireless MAN, based on the IEEE 802.16 family of standards, is a solution that can offer wireless broadband Internet access to residences and businesses at a relatively low cost. The standard supports shared transfer rates up to 75 Mbps from a single base station, which

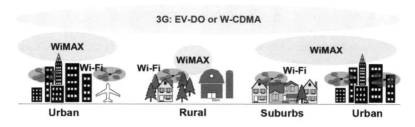

Figure 1.14 Complementing wireless broadband technologies

can offer broadband access without requiring a physical 'last-mile' connection from the end-user to a service provider. Service delivery to end clients is likely to be roughly 300 kbps for residences and 2 Mbps for businesses.

One of the promises of WiMAX is that it could offer the solution to what is sometimes called the 'last-mile' problem, referring to the expense and time of connecting individual homes and offices to trunk lines for communications. WiMAX promises a wireless access range of up to 31 miles, compared with 300 feet for Wi-Fi and 30 feet for Bluetooth.

To appreciate what WiMAX brings to the table we need to under-stand what additional features it provides over existing technologies. Existing broadband wireless access technologies that are closest to WiMAX with respect to service features are Wi-Fi and third-generation mobile. Let us first examine these three closely.

Wi-Fi has become one of the most popular forms of wireless local area networking, thanks to its open standard, high speed and ability to handle network interference. However, the popularity of Wi-Fi has exposed its primary limitation – range. The wireless technology can only serve signals in a 'hotspot' with a typical reach of about 1000 feet (300 m) outside or 328 feet (100 m) indoors, due to interference.

WiMAX is similar to the wireless standard known as Wi-Fi, but on a much larger scale and at faster speeds. A nomadic version would keep WiMAX-enabled devices connected over large areas, much like today's cell phones.

While Wi-Fi typically provides local network access for around a few hundred feet with speeds of up to 54 Mbps, a single WiMAX antenna is expected to have a range of up to 40 miles with speeds of 70 Mbps or more. As such, WiMAX can bring the underlying Internet connection needed to service local Wi-Fi networks.

'Wi-Fi does not provide ubiquitous broadband while WiMAX does'.

3G Cellular is long-range, mobile, reliable, plug-n-play, secure and private, with manageable data rates, but data applications are too expensive – as much as 10 times the cost of using similar wireline services.

Also, the biggest problem with 3G Cellular is the usage cost, which is due to factors like totally revamping the infrastructure and high licence fees.

'3G will not be affordable while WiMAX will be' (Figure 1.15).

Why WiMAX?

WiMAX can satisfy a variety of access needs. Potential applications include extending broadband capabilities to bring them closer to subscribers, filling gaps in cable, DSL and T1 services, Wi-Fi and cellular backhaul, providing last-100 m access from fibre to the curb and giving service providers another cost-effective option for supporting broadband services (Figure 1.16).

As WiMAX can support very high bandwidth solutions where large spectrum deployments (i.e. >10 MHz) are desired, it can leverage existing infrastructure, keeping costs down while delivering the bandwidth needed to support a full range of high-value, multimedia services. Further, WiMAX can help service providers meet many of the challenges they face due to increasing customer demands without discarding their existing infrastructure investments because it has the ability to seamlessly interoperate across various network types.

WiMAX can provide wide area coverage and quality of service capabilities for applications ranging from real-time delay-sensitive voice-over-IP (VoIP) to real-time streaming video and non-real-time downloads, ensuring that subscribers obtain the performance they expect for all types of communications.

WiMAX, which is an IP-based wireless broadband technology, can be integrated into both wide-area third-generation (3G) mobile and wireless and wireline networks, allowing it to become part of a seamless anytime, anywhere broadband access solution.

Ultimately, WiMAX is intended to serve as the next step in the evolution of 3G mobile phones, via a potential combination of

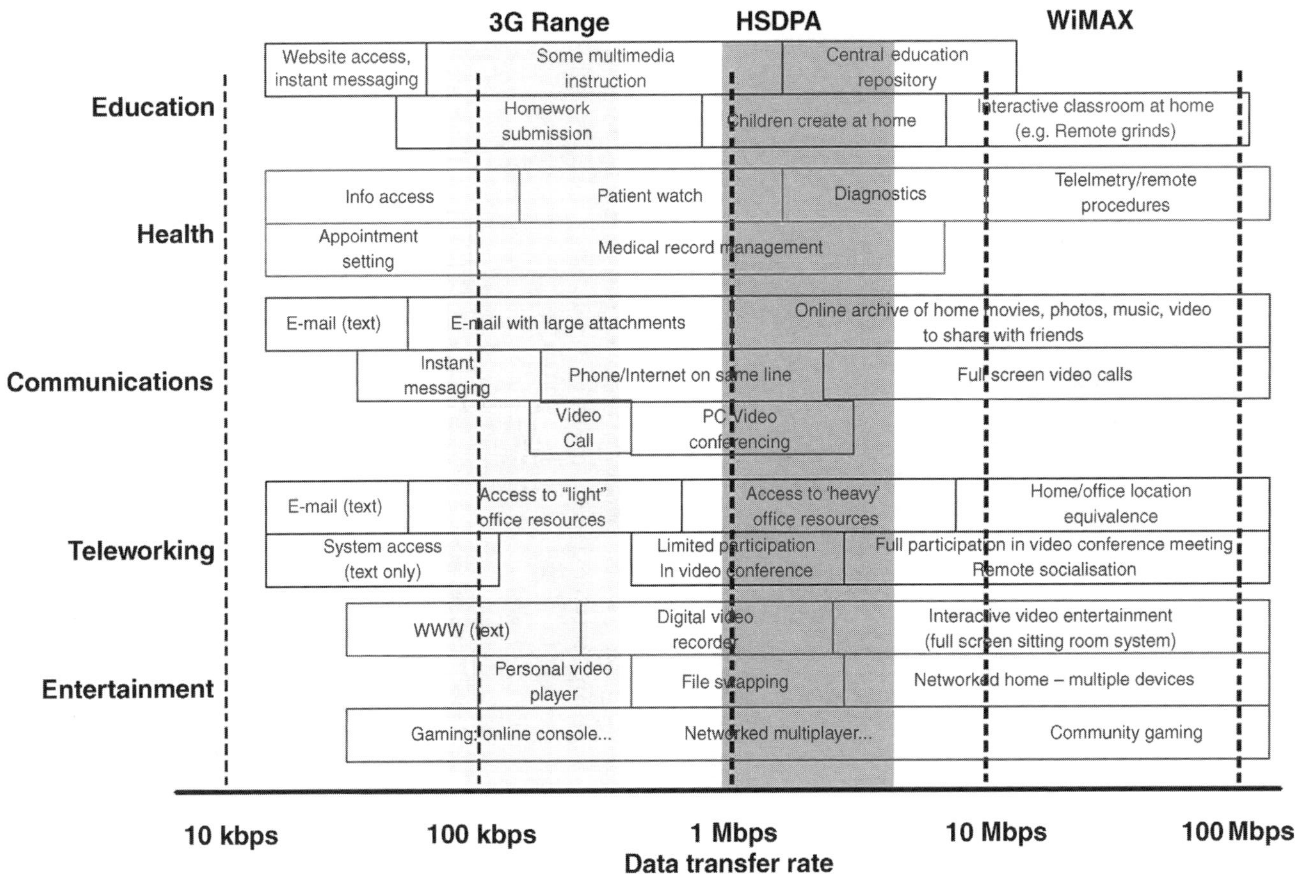

Figure 1.15 Applications using BWA

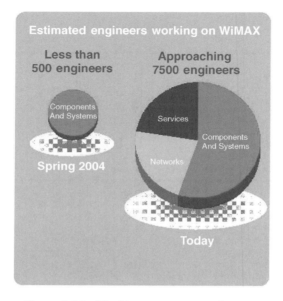

Figure 1.16   The big opportunity, July 2005

WiMAX and CDMA standards called 4G. Developed markets such as South Korea and Japan, where teledensity is over 100 %, are leading the charge in deploying this next generation of broadband wireless technologies (Figure 1.17).

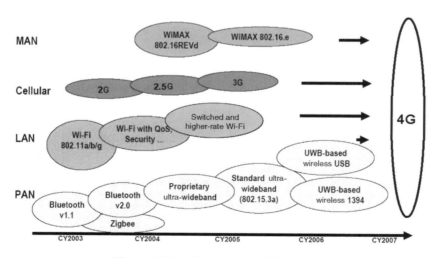

Figure 1.17   4G, just around the corner

1.7   WHAT NEXT? 4G

Every generation of mobile technology has a lifespan of about 10 years. Now, with the introduction of the third generation of mobile technology over the last couple of years, it is important to start designing future systems. There are still many issues that the third generation of mobile still does not solve, for example international standardization and unification of mobile and wireless technology. International standardization must be the primary objective for the telecommunications industry over the next 10 years.

It is always going to be a difficult task to try and predict, with some accuracy, what system and services will be provided in future. It would be ideal to have an integrated technology package and a single standard network called 4G. The main question is to ask whether the next generation of mobile systems will create a paradigm shift in how technology is perceived and used, or whether it will be a combination of existing networks built on top of existing platforms.

Wireless LANs have great potential in their environments and, although the technology has been around for a number of decades, it has only been in the last couple of years that the technology has accelerated. It now provides high bit rate (54 Mbps) and is relatively cheap to use.

The wired Internet has also seen a dramatic improvement in a number of areas. Better switching and routing techniques have realised the dream of a Giga-bit network. The introduction of flat rate charges has also greatly increased the use of the Internet.

The third generation of cellular phones has now increased the data rate and the number of services offered. It also allows the user to be charged by the amount of data transferred across the network and not the time spent in transferring the data.

Other mobile technologies such as Bluetooth and infrared transmission, allow the user to set up *ad hoc* wireless LANs, connect to wired LANs and create virtual private networks (VPNs).

There is no single technology that can provide the consumer with all the desired applications, but integrated of mobile and wireless systems could complement each other. All these technologies being available, it is quite reasonable to consider that a fourth generation of mobile system will give us the ability to integrate these technologies even though they work on different platforms.

As we have different techniques for communicating, an integrated system will have the ability to utilize all their advantages, so shortcomings of an individual technology will be complemented by other systems. Also, the big telecom companies have invested heavily in the new 3G network. At the same time there are many early adopters that have started building hotspots and Metrozones. The existing investments in WLAN and 3G technology will have a major impact on development of the future generation. Taking these aspects into consideration, the next generation of mobile systems, with a combination of existing networks built on top of existing platforms, seems a more realistic vision for the future.

The next generation of mobile and wireless systems will more than likely converge and integrate architectural, network and application levels. The next-generation network needs to be all IP in order to deal efficiently with the traffic. The unification of 3G, wireless LANs and DSL systems, as well as other access technologies, will enable users to access common services using TCP/IP running over them.

2

WiMAX in Depth

The growing demand for broadband services on a global scale is incontestable. Businesses, public institutions and private users regard it as an enabling technology and it has become a requirement for delivering communications services in the Information Age. The desire for bandwidth-intensive Internet access and other voice and data services has never been greater across all geographies and market segments, despite the air of uncertainty in the global telecommunications industry.

The DSL market, based on a variety of wireline infrastructures, has succeeded in reaching millions of business and private subscribers and continues on a rapid growth curve. In last-mile markets, where traditional cable or copper infrastructures are saturated, outdated or simply out of reach, supplying the quick roll-out of infrastructure to the last mile has become a difficult and expensive challenge for carriers who cannot possibly keep pace with demand.

This has brought about a situation wherein subscribers living in developed areas with broadband-ready infrastructure can enjoy all the benefits of DSL services while those who do not require another technology solution. Broadband Wireless Access (BWA) technology fills this void admirably, providing highly efficient and cost-effective access services for millions of subscribers who would otherwise be left out of the loop (Figure 2.1).

2.1 WiMAX: BROADBAND WIRELESS ACCESS

Broadband wireless is a technology that promises high-speed connection over the air. It uses radio waves to transmit and receive data directly to

The Business of WiMAX Deepak Pareek
© 2006 John Wiley & Sons, Ltd

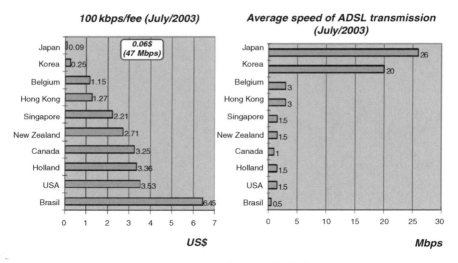

Figure 2.1 Cost and speed of access for different countries

and from the potential users whenever they want it. Technologies such as 3G, Wi-Fi, WiMAX and UWB work together to meet unique customer needs. BWA is a point-to-multipoint system which is made up of base station and subscriber equipment. Instead of using the physical connection between the base station and the subscriber, the base station uses an outdoor antenna to send and receive high-speed data and voice-to-subscriber equipment. This technology reduces the need for wireline infrastructure and provides a flexible and cost-effective last-mile solution (Table 2.1).

In current commercial deployments, broadband wireless networks deliver more bandwidth than traditional copper cables and exhibit a clear economical advantage over wireline alternatives in the last mile.

BWA offers an effective, complementary solution to wireline broadband, which has become globally recognized by a high percentage of the population. Technological improvements in the broadband wireless arena have been rapid and significant in recent years, offering operators greater performance and flexibility in their deployments while reducing their investment risks and ongoing operating expenses. Also, these technologies have been used for both commercial and residential application to solve connectivity needs. BWA thrives on an interoperability standard like any other segment of the information technology industry. A number of new wireless access technologies, and specifically the new WiMAX standard, fit this agenda perfectly.

Table 2.1 BWA development

2000	2001	2002	2003	2004	2005
Off-the-shelf 802.11 and proprietary solutions		Proprietary 70 + OEMs with equipment solutions		Standard-based solutions	
Spectrum: • Licence exempt 2.4 GHz • Lice333ned MMDS and LMDS		Spectrum: • Licence exempt 2.4 and $5\times$ GHz • Licensed 2.5, 3.5, 10.5 GHz		Harmonized IEEE 802.16a and ETSI HiperMAN standards Spectrum: <11 GHz licensed and licence-exempt	
Data rate: 2–11 Mbps peak Chip sets: use 802.11b, DOCSIS, proprietary		Data rate: 6–54 Mbps peak Chip sets: custom MAC, $802.11 \times$ PHY, DOCSIS Air interface: • OFDM and CDMA approaches • Proprietary antenna techniques (beam forming, MIMO)		Data rate: up to 75 Mbps peak Chip sets: volume 802.16a silicon Standard air interface: interoperable, carrier-class	

WiMAX is a wireless MAN technology that provides broadband wireless connectivity to fixed, portable and nomadic users. These powerful OFDM and NLOS technologies can potentially be used to provide backhaul to cellular networks, 802.11 hotspots and WLANs to the Internet, provide campus connectivity or significantly enhance the performance of public Wireless Fidelity (WiFi) hotspots by increasing the throughput in the backhaul network and by making it easier and more economical to deploy WiFi access points. WiMAX is also developing a 'mobile' standard which is not compatible with the fixed-based solution. The 'mobile' standard theoretically means that WiMAX can provide broadband wireless access in a vehicular environment. It provides up to 50 km of service area range, allows users to obtain broadband connectivity without needing direct line-of-sight with the base station, and provides total data rates of hundreds of Mbps per base station – a sufficient amount of bandwidth to simultaneously support hundreds of businesses with T1/E1-type connectivity and thousands of homes with DSL-type connectivity with a single base station. Further, WiMAX offers last-mile connection to remote un-served customers, as many are outside the range of DSL or broadband cable with WiMAX these barriers can be lifted and new customers can be captured.

WiMAX is faster to deploy, easier to scale and more flexible; thus, it gives an alternative service to customers who are not satisfied with their wired broadband. WiMAX is revolutionizing the broadband wireless world, enabling the formation of a global mass-market wireless industry (Figure 2.2).

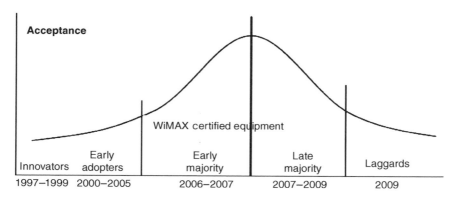

Figure 2.2   WiMAX chasm – acceptance

2.2 WiMAX REVOLUTION

Historically, many operators worldwide have used broadband wireless technologies, namely point-to-point radios, as a proven, service provider-class method of connecting long-haul networks. While point-to-point technologies have also been used for access in isolated cases (with mixed results), point-to-multipoint technologies have long been recognized as the 'holy grail' for service providers because of their ability to provide broadband services over large geographic areas with greater flexibility and improved economies of scale. The industry has suffered from limited deployments, however, due to the high cost and low functionality of the early generation of broadband wireless systems.

In addition, a lack of healthy competition caused by various factors has contributed to the industry's malaise. In addition, with frequency regulations varying from country to country, equipment manufacturers have used only proprietary air interface technologies. These and other factors have precluded the broadband wireless market from benefiting from the economies of scale that other technologies enjoy from open standards.

WiMAX addresses the deficiencies (as indicated below) of previous BWA initiatives:

- interoperability;
- cost of base stations and CPE;
- shared bandwidth up to 100 Mbps;
- line-of-sight not required;
- coverage 3–5 miles, more like cellular;
- licensed and unlicensed spectrum;
- many DOCSIS-like features including QoS;
- milestone to 'broadband everywhere'.

What differentiates WiMAX from earlier BWA iterations is standardization. Chipsets are currently custom-built for each BWA vendor, adding time and cost to the process. Intel and other chipmakers would like to bring scale to the market and BWA vendors are always interested in less expensive chipsets. Similar to the way that the WiFi Alliance enforced standards compliance among vendor members, the WiMAX Forum plans to do the same among its members. Compliance results in interoperability, which in turn results in plug-and-play products. In the years ahead, WiMAX vendors will no longer have to provide   end-to-end solutions; they can specialize in base stations or in wireless modems. Specialization will result in competitive pricing and value-added innovations.

WiMAX is the most important of the host of wireless standards emerging from IEEE. Its impact will owe much to WiFi, which has created interest in and market acceptance of wireless networking that will enable WiMAX to flourish in the mainstream, not least by attracting Intel into the sector. However, its effect on the world of business and consumer Internet and wireless access will be far more profound.

WiMAX has become one of the most talked-about and eagerly anticipated developments among wireless telecom carriers and equipment providers throughout the world. The reason is straightforward: WiMAX responds to the real challenges faced by carriers in deploying wireless networks.

WiMAX will enable service providers to economically achieve what they have long been striving for – to deliver broadband data, voice and video services to residential and business clients worldwide. Expected to provide a cost-effective last-mile solution in delivering high-speed BWA to a variety of communities while accommodating several service needs, WiMAX gives hope to both current and emerging broadband markets. WiMAX is the most significant technology to date in making wireless access ubiquitous and, as more spectrum is opened up, in creating a major shake-up of the wireless and mobile communications sector.

WiMAX is well suited to providing varying populations with the precise amount of bandwidth they need. With an acute awareness of prior BWA industry pitfalls, namely the abundance of proprietary technologies failing to find confluence, key business players are addressing the interoperability issue throughout the development of WiMAX. The use of standards-based, interoperable equipment will enable bulk-volume production, spark competition and maintenance of lower costs, and lead to economies of scale. Furthermore, the proliferation of WiMAX should simplify BWA deployment. Without the costly, cumbersome task of laying down wire and cable, densely populated, developing countries are more likely to quickly reap the benefits of high-speed Internet connectivity.

The case for WiMAX

The economic case for broadband wireless access networks is simple: wireless in many cases is the most effective medium for transporting data, video and voice traffic, and offers much higher bandwidth in the access network than existing wireline options with the exception of fibre.

The cost of running fibre point-to-point from every potential customer location to the central office or POP, installing electronics at both ends of each fibre and managing all of the fibre connections at the central office is prohibitive.

Broadband wireless access networks built upon WiMAX address the shortcomings of fibre solutions by using a point-to-multipoint topology instead of point-to-point in the access network, eliminating the cost of installation and the reoccurring operating cost due to the use of active electronic components such as regenerators, amplifiers and lasers within the outside plant, and reducing the number of network elements needed at the central office.

WiMAX is Different

If we compare the WiMAX phenomenon to earlier build-up for 3G technologies, one of the key differentiators is that carriers and end-users – not manufacturers – drive the need for WiMAX services. Indeed, the proliferation of broadband DSL and cable services has produced a mature end-user market that is now demanding high-speed services anytime, anywhere. Similarly, WiFi has given the world a taste for wireless broadband.

In essence, WiMAX is a set of standards focused on supporting high-speed IP communications across the wireless infrastructure, in concert with the direction many carriers are taking towards an all-IP wireline network. WiMAX has the potential to significantly change the world's telecommunications landscape. CLECs would be able to provide a real broadband alternative using their own infrastructure; ILECs would be able to deploy high-speed Internet access in regions where wired connections are not profitable; and WISPs using WiFi technologies would be able to extend their existing services.

WiMAX is intended to provide definitive IP standards for a carrier-class solution that can scale to support thousands of users with a single base station, and provide differentiated service levels. By enabling IP standards-based products with fewer variants and larger volume production, WiMAX should drive down the cost of equipment and make broadband wireless a real alternative to wireline technologies. Soon a single base station sector will provide a sufficiently high data rate to simultaneously support more than 60 businesses with T1-type connectivity and hundreds of homes with DSL-type connectivity.

Much effort has gone into making the wireless technology very robust and flexible, so it will work well in a range of different environments around the world. This was a major area of work in the development of the 802.16a version. For example, it can withstand the effects of multiple radio reflections (or echoes) from buildings and other obstacles in the transmission path – a major problem in built-up environments.

Different channel sizes and methods of providing two-way communications are supported so that the technology can accommodate different national regulatory and technical requirements. Importantly, WiMAX supports smart antenna systems, which are rapidly becoming less expensive and are very effective in reducing the effects of radio interference and the wireless power needed. This is done by using four antennas at the base station instead of just one. Each of the four antennas transmits and receives the same data signal, but at slightly different times. By clever signal processing, the best signal can always be extracted. To achieve the same performance with a single antenna, more wireless power would be needed, increasing costs and the problems of interference and cell planning.

A further bonus of WiMAX is that it supports mesh networks. This means that WiMAX-enabled devices can act as relays, passing signals from one device to another until they reach a WiMAX base station from which they can enter the wired Internet. Relaying like this greatly extends the potential range of an access point, and allows networks to grow in an organic fashion (Figure 2.3).

2.3   WiMAX: WORKING

WiMAX has been designed to address challenges associated with traditional wired and wireless access deployments. A WiMAX network has a number of base stations and associated antennas communicating by wireless to a much larger number of client devices (or subscriber stations).

The WiMAX MAN is schematically similar to the point-to-multipoint layout of a cellular network. It revolves around strategically positioned, highly elevated base stations that beam signals to CPE within their radii. The original 802.16 specification paved the way for fixed wireless-access coverage, which requires a mounted outdoor antenna at the customer's access point. However, this fixed coverage will soon evolve to incorporate indoor antennas before altogether segueing into an even more

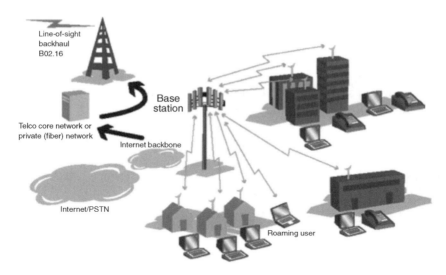

Figure 2.3 WiMAX working

significant development: the 802.16e 'mobility' extension. Where fixed wireless-access coverage CPE can only communicate with their respective base station, this revision would enable seamless communication from station to station.

A WiMAX base station is connected to public networks using optical fibre, cable, microwave link or any other high-speed point-to-point connectivity, referred to as a backhaul. In a few cases, like mesh networks, a point-to-multipoint WiMAX link to other base stations is used as a backhaul. Ideally WiMAX should use point-to-point antennas as a backhaul to connect aggregate subscriber sites to each other and to base stations across long distances.

The base station serves subscriber stations (also called customer premise equipment) using non-line-of-sight or line-of-sight point-to-multipoint connectivity referred to as 'last mile'. Ideally WiMAX should use non-line-of-sight point-to-multipoint antennas to connect residential or business subscribers to the base station.

The subscriber station typically serves a building (business or residence) using wired or wireless LAN. Initially they are generally small, building-mounted antenna/transceiver systems to which in-building LANs (such as WLANs) are connected. However, future clients depending on the frequency bands used will often be integrated into end-user devices, such as notebook PCs and, eventually, mobile devices, such as PDAs and smart phones.

WiMAX: Key Capabilities

Centrally coordinated architecture:

- high-end security, encryption and service authentication;
- no *ad-hoc* PP client communication is possible;
- robust radio interface that works in NLOS conditions;
- OFDM PHY supports indoor, self installation by end users.

High-speed IP services:

- optimized to deliver 110 Mbps (net) services (in 3.5 MHz);
- Up to 3550 Mbps (net) with large channels (1420 MHz).

Second generation IP QoS:

- hierarchical QoS supports real-time and grant-based service delivery (not just best effort!).

A low delay radio interface:

- enables latency and jitter sensitive applications (VoIP, Internet Gaming, etc.).

2.4 WiMAX: BUILDING BLOCKS

Typically, a WiMAX system consists of two parts: a WiMAX base station and a WiMAX receiver, also referred as customer premise equipment (CPE). While the backhaul connects the system to the core network it is not the integrated part of WiMAX system as such.

WiMAX Base Station

A WiMAX base station consists of indoor electronics and a WiMAX tower. Typically, a base station can cover up to a radius of 6 miles (theoretically, a base station can cover up to a 50 km radius or 30 miles; however, practical considerations limit it to about 10 km or 6 miles). Any wireless node within the coverage area would be able to access the Internet.

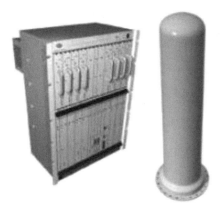

Figure 2.4   WiMAX base station

The WiMAX base stations would use the MAC layer defined in the standard – a common interface that makes the networks interoperable – and would allocate uplink and downlink bandwidth to subscribers according to their needs, on an essentially real-time basis.

Each base station provides wireless coverage over an area called a cell. The maximum radius of a cell is theoretically 50 km (depending on the frequency band chosen); however, typical deployments will use cells of radii from 3 to 10 km.

As with conventional cellular mobile networks, the base-station antennas can be omnidirectional, giving a circular cell shape, or directional to give a range of linear or sectoral shapes for point-to-point use or for increasing the network's capacity by effectively dividing large cells into several smaller sectoral areas (Figure 2.4).

WiMAX Receiver

A WiMAX receiver may have a separate antenna (i.e. receiver electronics and antenna are separate modules) or could be a stand-alone box or a PCMCIA card that sits in your laptop or computer. Access to a WiMAX base station is similar to accessing a wireless access point in a WiFi network, but the coverage is greater.

So far one of the biggest deterrents to the widespread acceptance of BWA has been the cost of CPE. This is not only the cost of the CPE itself, but also the installation cost. Historically, proprietary BWA systems have been predominantly line-of-site, requiring highly skilled labour to install and 'turn up' a customer device. The concept of a

Figure 2.5   WiMAX receiver

self-installed CPE has been the Holy Grail for BWA from the beginning. With the advent of WiMAX this issue seems to be resolving (Figure 2.5).

Backhaul

Backhaul refers both to the connection from the access point back to the provider and to the connection from the provider to the core network. A backhaul can deploy any technology and media provided it connects the system to the backbone. In most of the WiMAX deployment scenarios, it is also possible to connect several base stations to one another using high-speed backhaul microwave links. This would also allow for roaming by a WiMAX subscriber from one base station coverage area to another, similar to the roaming enabled by cell phones (Figure 2.6).

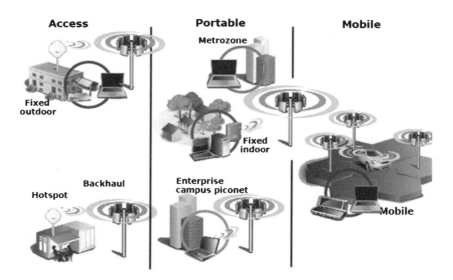

Figure 2.6   WiMAX technology

2.5   WiMAX TECHNOLOGY

802.16 was originally designed to provide a flexible, cost-effective, standards-based last-mile broadband connectivity to fill in the broadband coverage gaps that are not currently served by 'wired' solutions such as cables or DSL, the evolved versions of the standard are aiming to create new forms of broadband services both with high speed and mobility.

WiMAX is a technology based on the IEEE 802.16 specifications to enable the delivery of last-mile wireless broadband access as an alternative to cable and DSL. WiMAX will provide fixed, nomadic, portable and eventually mobile wireless broadband connectivity without the need for direct line-of-sight with a base station. The design of WiMAX network is based on the following major principles:

- Spectrum – able to be deployed in both licensed and unlicensed spectra.
- Topology – supports different radio access network (RAN) topologies.
- interworking – independent RAN architecture to enable seamless integration and interworking with WiFi, 3GPP and 3GPP2 networks and existing IP operator core network (e.g. DSL, cable, 3G) via IP-based interfaces which are not operator-domain specific.
- IP connectivity – supports a mix of IPv4 and IPv6 network interconnects in clients and application servers.
- Mobility management – possibility to extend the fixed access to mobility and broadband multimedia services delivery.

WiMAX created the 10–66 GHz technical working group as a testing mission in the IEEE. WiMAX has defined two MAC system profiles the basic ATM and the basic IP. They have also defined two primary PHY system profiles, the 25 MHz-wide channel for use (typically for US deployments) in the 10–66 GHz range, and the 28 MHz wide channel for use (typically European deployments) in the 10–66 GHz range.

WiMAX expanded their work to include the 802.16a standard in terms of addressing testing and conformance issues. The 2–11 GHz technical working group has the mandate of creating testing and conformance documents as contributions to IEEE and ETSI standards bodies in support of the IEEE 802.16a and ETSI HiperMan standards.

The WiMAX technical working group is defining MAC and PHY system profiles for IEEE 802.16a and HiperMan standards. The MAC profile includes an IP-based version for both wireless MAN (licensed) and wireless HUMAN (licence-exempt).

IEEE Standard 802.16 was designed to evolve as a set of air interfaces standards for WMAN based on a common MAC protocol but with physical layer specifications dependent on the spectrum of use and the associated regulations. The IEEE 802.16 working group designed a flexible MAC layer and accompanying physical layer (PHY) for 10–66 GHz (Figure 2.7).

Architecture

The architecture and usage of WiMAX have a two-stage evolution: fixed access is initially combined with portability and then scaled up to

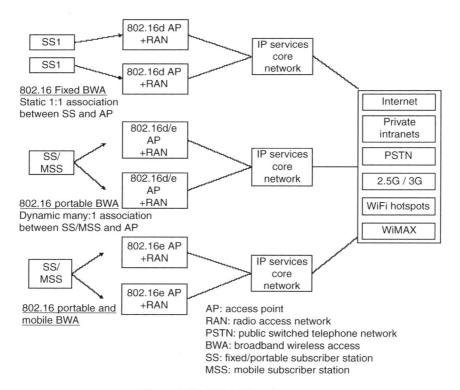

Figure 2.7   WiMAX architecture

evolve to full mobility. The framework is based on several core principles:

- support for different RAN topologies;
- well-defined interfaces to enable 802.16 RAN architecture independence while enabling seamless integration and interworking with WiFi, 3GPP3 and 3GPP2 networks;
- leverage and open, IETF-defined IP technologies to build scalable all-IP 802.16 access networks using common off the shelf (COTS) equipment;
- support for IPv4 and IPv6 clients and application servers, recommending use of IPv6 in the infrastructure;
- functional extensibility to support future migration to full mobility and delivery of rich broadband multimedia.

The architecture framework is based on the following requirements:

Applicability

The architecture will be applicable to licensed and licence-exempt 802.16 deployments. The architecture, especially the RAN, will be suitable for adoption by all incumbent operator types, examples of which were listed earlier. The architecture will lend itself to integration with an existing IP operator core network (e.g. DSL, cable or 3G) via interfaces that are IP-based and not operator-domain specific. This permits reuse of mobile client software across operator domains.

Provisioning and Management

The architecture will accommodate a variety of online and offline client provisioning, enrolment and management schemes based on open, broadly deployable industry standards.

IP Connectivity and Services

The architecture will support a mix of IPv4 and IPv6 network interconnects and communication endpoints and a variety of standard IP context management schemes. The architecture will support a broad range of TCP and UDP real-time and non-real-time applications.

Security

The architecture will support subscriber station (SS) authorization, strong bilateral user authentication based on a variety of authentication mechanisms, such as username/password, X.509 certificates, subscriber identity module (SIM), universal SIM (USIM) and removable user identity module (RUIM), and provide services such as data integrity, data replay protection, data confidentiality and non-repudiation using the maximum key lengths permissible under global export regulations.

Mobility Management

The architecture will scale from fixed access to fully mobile operation scenarios with scalable infrastructure evolution, eventually supporting low latency (<100 ms) and virtually zero packet loss handovers at mobility speeds of 120 km/h or higher. _plan_

MAC Layer

The IEEE 802.16 MAC was designed for point-to-multipoint broadband wireless access applications. The MAC was developed by Task Group 1 along with the original 10–66 GHz PHY. The original design of MAC was flexible enough to support, with an extension, all other projects of the IEEE 802.16. The design addresses the need for very high bit rates, both uplink and downlink. The MAC must accommodate both continuous and burst traffic to support the variety of services required by multiple end users. These services are varied in their nature and include legacy time-division multiplex (TDM) voice and data, IP connectivity and packetized VoIP.

The 802.16 MAC provides the option of allowing a smart subscriber station to manage its bandwidth allocation among its users, and this is due to the MAC characteristic that offers the choice of conceding bandwidth to a subscriber station rather than to the individual connection it supports.

The 802.16 MAC is adaptable and flexible, and it supports several multiplexing and duplexing schemes. The MAC consists of three sublayers: the service-specific convergence sublayer (SSCS), the MAC common part sublayer (CPS) and the privacy sublayer. In the 802.16 MAC protocol, convergence sublayers are used to map the transport-layer-specific traffic to a MAC that is flexible enough to efficiently carry that traffic type.

The 802.16 MAC is designed for point-to-multipoint (PMP) applications and is based on collision sense multiple access with collision avoidance (CSMA/CA). The 802.16 AP MAC manages UL and DL resources including transmit and receive scheduling. The MAC incorporates several features suitable for a broad range of applications at different mobility rates, such as the following:

- four service classes, unsolicited grant service (UGS), real-time polling service (rtPS), non-real-time polling service (nrtPS) and best effort (BE);
- header suppression, packing and fragmentation for efficient use of spectrum;
- privacy key management (PKM) for MAC layer security. PKM version 2 incorporates support for extensible authentication protocol (EAP);
- broadcast and multicast support;
- manageability primitives;
- high-speed handover and mobility management primitives;
- three power management levels, normal operation, sleep and idle (with paging support).

These features combined with the inherent benefits of scalable OFDMA make 802.16 suitable for high-speed data and bursty or isochronous IP multimedia applications (Table 2.2).

Physical (PHY) Layer

The 10–66 GHz PHY assumes line-of-sight propagation with no considerable concern over multipath propagation. The PHY layer contains several forms of modulation and multiplexing to support different frequency range and application.

The IEEE 802.16 standard was originally written to support several physical medium interfaces, and it is expected that it will continue to

Table 2.2   802.16 MAC features

Feature	Benefits
TDM/TDMA scheduled uplink/downlink frames.	• Efficient bandwidth usage
Scalable from 1 to hundreds of subscribers	• Allows cost-effective deployments by supporting enough subscribers to deliver a robust business case
Connection-oriented	• Per connection QoS • Faster packet routing and forwarding
QoS support continuous grant real-time variable bit rate non-real-time variable bit rate best effort	• Low latency for delay-sensitive services (TDM voice, VoIP) • Optimal transport for VBR traffic (e.g. video) • Data prioritization
Automatic retransmission request (ARQ)	• Improves end-to-end performance by hiding RF layer-induced errors from upper layer protocols
Support for adaptive modulation	• Enables highest data rates allowed by channel conditions, improving system capacity
Security and encryption (triple DES)	• Protects user privacy
Automatic power control	• Enables cellular deployments by minimizing self interference

develop and extend to support other PHY specifications. Hence, the modular nature of the standard is helpful in this aspect. For example, the very first version of the standard only supported single carrier modulation. Since that time, orthogonal frequency division multiplexing (OFDM) has been added (Table 2.3; Figure 2.8).

2.6   WiMAX STANDARDS

The IEEE 802.16, the 'Air Interface for Fixed Broadband Wireless Access Systems', also known as the IEEE WirelessMAN air interface, is an emerging suite of standards for fixed, portable and mobile BWA in MAN.

These standards are issued by IEEE 802.16 work group that originally covered the wireless local loop (WLL) technologies in the 10–66 GHz radio spectrum, which were later extended through amendment projects to include both licensed and unlicensed spectra from 2 to 11 GHz.

Although the term WiMAX is only a few years old, 802.16 has been around since the late 1990s, first with the adoption of the 802.16

Table 2.3   802.16 PHY features

Feature	Benefits
256-point FFT OFDM waveform	Built-in support for addressing multipath in outdoor LOS and NLOS environments
Adaptive modulation and variable error correction encoding per RF burst	Ensures a robust RF link while maximizing the number of bps for each subscriber unit
TDD and FDD duplexing support	Addresses varying worldwide regulations where one or both may be allowed
Flexible Channel sizes (e.g. 3.5, 5, 10 MHz.)	Provides the flexibility necessary to operate in many different frequency bands with varying channel requirements around the world
Designed to support smart antenna systems	Smart antennas are fast becoming more affordable, and as these costs come down their ability to suppress interference and increase system gain will become important to BWA deployments

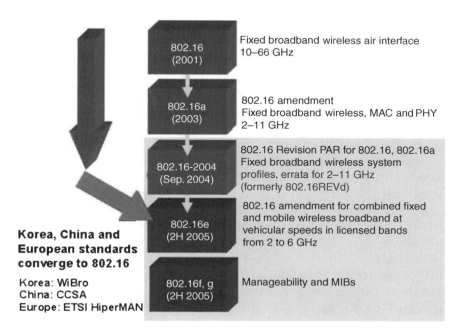

Figure 2.8   WiMAX standards

standard (10–66 GHz) and then with 802.16a (2–11 GHz). Although the work on IEEE 802.16 standard started in 1999, it was only during 2003 that the standard received wide attention when the IEEE 802.16a standard was ratified in January.

Developing a new technical standard in the telecommunications field is never an easy process, but the growth of wireless LANs and WiFi hotspots has shown that there is a huge demand for wireless services that allow people to connect to the Internet and business networks from anywhere at any time. That has provided equipment makers and service providers with the incentive to cooperate on developing standards for WiMAX in order to grow the market. Standardization efforts have been underway for almost 6 years and momentum is building for multi-vendor interoperability.

The IEEE 802.16 standards for BWA provide the possibility for interoperability between equipment from different vendors, which is in contrast to the previous BWA industry, where proprietary products with high prices are dominant in the market. The IEEE 802.16 standards have received wide support from major equipment manufacturers like Intel and Nokia.

Furthermore, a nonprofit organization called WiMAX Forum was formed in 2001, with the aim of harmonizing standards, and testing and certifying interoperability between equipment from different manufacturers. Systems conforming to the test specifications will receive a 'WiMAX certified' label. This standardized solution is expected to bring beneficial economical effects such as mass production and cost reduction.

The stage is set for a paradigm-shift in the communications industry that could well result in a completely new equipment deployment cycle, firmly grounded in the wide-based adoption of Ethernet technologies. This BWA network architecture promises to become a significant means of delivering bundled voice, data and video services over a single network.

It is often thought that WiMAX is one homogenous technology when in fact it is a trade name for a group of IEEE wireless standards. In that respect, WiMAX and WiFi are analogous. WiFi is not a standard, but a trade name that can be applied to a series of 802.11 IEEE standards, including 802.11b, 802.11a and 802.11g. It is assumed that the term WiFi will be applied to 802.11n once that standard is ratified.

The WiMAX umbrella currently includes 802.16-2004 and 802.16e. 802.16-2004 utilizes OFDM to serve multiple users in a time division fashion in a sort of a round-robin technique, but done extremely quickly so that users have the perception that they are always transmitting/receiving. 802.16e utilizes OFDMA and can serve multiple users simultaneously by allocating sets of 'tones' to each user (Table 2.4).

Table 2.4   IEEE 802.16 standards

Standard	Classification	Remarks	Status
802.16	Air interface	WirelessMAN^{TM} standard (air interface for fixed broadband wireless access systems) for wireless metropolitan area networks	Published April 2002
802.16a	Air interface	Amendment to 802.16; purpose is to expand the scope to licensed and license-exempt bands from 2 to 11 GHz	Published April 2003
802.16c	Air interface	Amendment to 802.16; purpose is to develop 10–66 GHz system profiles to aid interoperability specifications	Published January 2003
802.16REVd	Air interface	Converted from 802.16d, now published as the most recent update to the standard	Approved as 802.16-2004 in June 2004
802.16.2	Coexistence	Recommended practice on coexistence of broadband wireless access systems for 10–66 GHz	Published September 2001; now replaced by 802.16.2-2004
802.16.2a	Coexistence	Amendment to 802.16.2; purpose is to expand scope to include licensed bands from 2 to 11 GHz and to enhance the recommendations regarding point-to-point systems	Subsequently converted and published as 802.16.2-2004 in March 2004
802.16/Conf01 Conformance		Conformance01 PICS for 10–66 GHz	Published August 2003
802.16/Conf02 Conformance		Test suite structure and test purposes for 10–66 GHz	Published February 2004
802.16/Conf03 Conformance		10–66 GHz radio conformance tests	Approved May 2004
802.16/Conf04	Conformance	PICS for <11 GHz	Pending

IEEE 802.16-2004

IEEE 802.16-2004 is a fixed wireless access technology, meaning that it is designed to serve as a wireless DSL replacement technology, to compete with the incumbent DSL or broadband cable providers or to provide basic voice and broadband access in under-served areas where no other access technology exists; examples include developing countries and rural areas in developed countries where running copper wire or cable does not make economic sense. 802.16-2004 is also a viable solution for wireless backhaul for WiFi access points or potentially for cellular networks, in particular if licensed spectrum is used. Finally, in certain configurations, WiMAX fixed can be used to provide much higher data rates and therefore be used as a T1 replacement option for high-value corporate subscribers.

Typically, the CPE consists of an outdoor unit (antenna etc.) and an indoor modem, meaning that a technician is required to connect a commercial or residential subscriber to the network. In certain instances, a self-installable indoor unit can be used, in particular when the subscriber is in relatively close proximity to the transmitting base station. The trend towards self-installable indoor units is likely to develop more noticeably in the next few years. As it does, the fixed wireless technology would introduce a degree of nomadic capability since the subscriber could travel with the CPE and use it in other fixed locations, e.g. office, hotel or coffee shop. Additionally, self-installable CPEs should make 802.16-2004 more economically viable as much of the customer acquisition cost (installation, CPE) is reduced. Although it is technically feasible to design an 802.16-2004 data card, handheld devices with an embedded 802.16-2004 solution do not appear to be a top priority within the industry at this time.

The fixed version of the WiMAX standard was approved in June 2004, although interoperability testing will not begin until later in 2005. A project to fix bugs in the published standard is ongoing, and is expected to be completed in September 2005. Further, base station and CPE chipsets from the major vendors are just reaching the point where potential customers have been sampling them with the Intel Rosedale chipset sampling since September 2004 and Fujitsu announced its first WiMAX chipset early in 2005 (Figure 2.9).

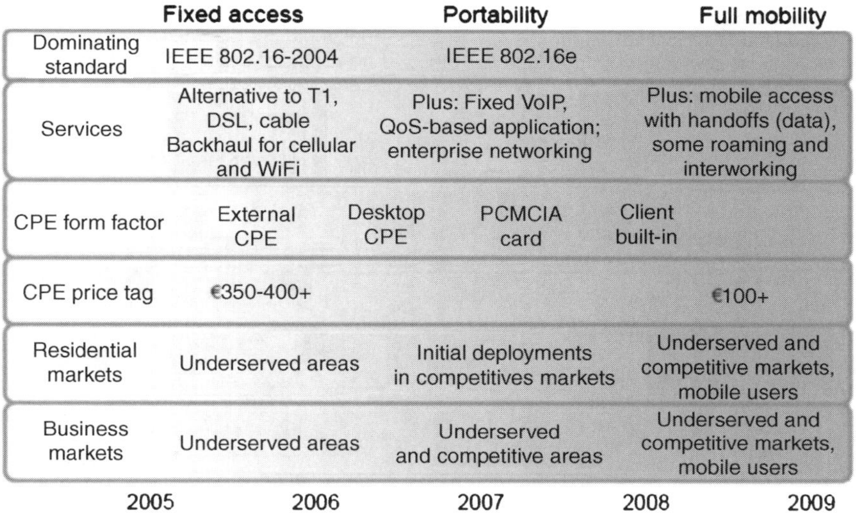

	Fixed access	Portability		Full mobility
Dominating standard	IEEE 802.16-2004	IEEE 802.16e		
Services	Alternative to T1, DSL, cable Backhaul for cellular and WiFi	Plus: Fixed VoIP, QoS-based application; enterprise networking		Plus: mobile access with handoffs (data), some roaming and interworking
CPE form factor	External CPE	Desktop CPE	PCMCIA card	Client built-in
CPE price tag	€350-400+			€100+
Residential markets	Underserved areas	Initial deployments in competitives markets		Underserved and competitive markets, mobile users
Business markets	Underserved areas	Underserved and competitive areas		Underserved and competitive markets, mobile users
2005	2006	2007	2008	2009

Figure 2.9   WiMAX standard – fixed to mobile

IEEE 802.16e

IEEE 802.16e is as yet an unpublished standard that is intended to offer a key feature that 802.16-2004 lacks – portability and eventually full-scale mobility. This standard requires a new hardware/software solution since it is not backward-compatible with 802.16-2004 – not necessarily a good thing for operators that are planning to deploy .16-2004 and then upgrade to .16e.

Another major difference between the .16-2004 and .16e standards is that the .16-2004 standard is based, in part, upon a number of proven, albeit proprietary, fixed wireless solutions, thus there is a good likelihood that the technology will achieve its stated performance targets. The .16e standard, on the other hand, tries to incorporate a wide variety of proposed technologies, some more proven than others. Since there has been only modest justification of proposed features on the basis of performance data, and the final composition of these technologies has not been completely determined, it is difficult to know whether a given feature will enhance performance.

From a timing perspective, the 802.16e standard was scheduled to be approved in mid-2005; however, it got delayed and was approved in last quarter of 2005. Several vendors are promising field and market

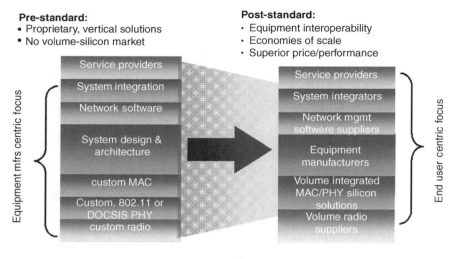

Figure 2.10 Benefits of standard

trials in early 2006, although, as discussed later in this paper, much work still remains to be done outside the standards body and it is therefore too early to tell when the technology will be ready for commercial deployments (Figure 2.10; Table 2.5).

Table 2.5 IEEE 802.16 timelines – basic features

	802.16	**802.16a**	**802.16e**
Completed	2001	2003	2005
Spectrum	10–66 GHz	2–11 GHz	2–6 GHz
Channel conditions	Line-of-sight Only	Non-line-of-sight	Non-line-of-sight
Bit rate	32–134 Mbps at 28 MHz channelization	Up to 75 Mbps at 20 MHz channelization	Up to 15 Mbps at 5 MHz channelization
Modulation	QPSK, 16QAM and 64QAM	OFDM 256 sub-carriers QPSK, 16QAM, 64QAM	Same as 802.16a
Mobility	Fixed	Fixed	Pedestrian mobility – regional roaming
Typical cell radius	1–3 miles	4–6 miles; Maximum range 30 miles based on tower height, antenna gain and transmit power	1–3 miles

2.7 WiMAX FORUM^{TM}

WiMAX Forum is an industry-led, non-profit corporation formed to promote and certify compatibility and interoperability of broadband wireless products. It was formed by equipment and component suppliers to support the IEEE 802.16 BWA system by helping to ensure the compatibility and interoperability of BWA equipment which will lead to lower cost through chip-level implementation.

Visionary companies, which feel that proprietary wireless access solutions are inappropriate for the wireless access network of the future, have spearheaded the development of WiMAX Forum Certified^{TM} products. The proprietary solutions are wanting because they generally lack video capabilities, deliver insufficient bandwidth, are overly complex and are more expensive due to a nonintegrated supply chain. As the move to Ethernet picks up steam, Ethernet-based broadband wireless access networks will eliminate the need for multiple conversions within the WAN and LAN between ATM and IP network elements.

WiMAX Forum is doing what WiFi Alliance has done for wireless LAN and IEEE 802.11. In 2003 Intel Corp became a major supporter of the WiMAX Forum. WiMAX Forum Certified^{TM} products adhere to the IEEE 802.16 standard and offer higher bandwidth, lower costs and broader service capabilities than most of the available proprietary solutions.

The WiMAX Forum is working very hard to set up a baseline protocol that allows equipment and devices from multiple vendors to interoperate, and also provides a choice of equipment and devices from different suppliers. Equipment vendors are focusing initially on developing comprehensive transport and access solutions, with the long-term objective of realizing a full-service standards-based solution for delivering data, video and voice over a single platform.

The WiMAX Forum has 300 members from equipment manufacturers, semiconductor suppliers, and services providers, and membership was recently opened for content providers. Some of the noted members are Alcatel, AT&T, Fujitsu, Intel, Nortel, Motorola, SBC and Siemens.

You should be aware of three additional facts about how WiMAX relates to official standards:

- First, broadly WiMAX is seen to be associated with the 802.16REVd.
- Second, the WiMAX Forum's profiles will also be compatible with the European HiperMAN standard.

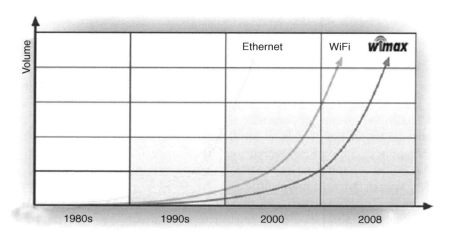

Figure 2.11   Forums – making technology succeed

- Finally, the WiMAX Forum also has its sights set on 802.16e, an additional amendment that is awaiting approval by the IEEE, (Figure 2.11).

Why WiMAX Forum™?

The purpose of the WiMAX Forum™ is to promote deployment of BWA networks by using a global standard (802.16) and certifying interoperability of products and technologies manufactured by member companies.

The WiMAX Forum™ will follow the IEEE tradition of being a clear, comprehensive and complete standard, designed for very high volume applications. All players will experience the most effective wireless infrastructure for broadband data services whether fixed, nomadic or eventually mobile. They will experience future-proof transport for data, video and voice applications, all through a simple global interoperability and certification process that will ensure interoperability (Table 2.6).

WiMAX Forum Certified™ is an open certification based on IEEE Standard 802.16d. The WiMAX Forum™ also plans to certify that products are compliant with the interoperability requirements set forth by The WiMAX Forum™ Technical Working Group. For network operators, this interoperability gives the operator more options, the

Table 2.6   WiMAX Forum – mission and principles

Mission	Principles
To promote deployment of BWA by using a global standard and certifying interoperability of products and technologies.	Support IEEE 802.16, 2–66 GHz • Propose access profiles for the IEEE 802.16 standard • Guarantee known interoperability level • Promote IEEE 802.16 standard to achieve global acceptance • Open for everyone to participate

flexibility of deploying broadband wireless systems from multiple vendors and the knowledge that all products deployed, if certified, will interoperate seamlessly (Figure 2.12).

WiMAX Profiles

The 802.16 standard covers a lot of ground. When the IEEE approved it in 2001, it addressed wireless communications in the 10–66 GHz range and targeted line-of-sight, point-to-point applications. In January 2003, the IEEE approved an amendment, 802.16a, which expanded the scope to include the 2–11 GHz range and, more important, set the stage for point-to-multipoint, non-line-of-sight applications.

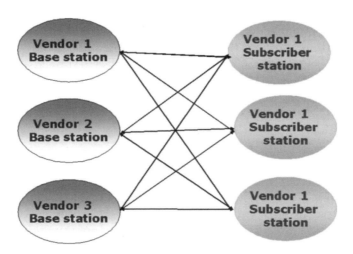

Figure 2.12   WiMAX forum – interoperability

The WiMAX Forum is focusing on 802.16a as of now and more specifically on one of the four PHYs that the standard allows. As 802.16 is very flexible, the only way to achieve interoperability is to narrow it down to a certain set of system options. Since 802.16-2004 addresses the entire sub-11 GHz frequency range, there is an inherent need for a number of different solutions or profiles to use the vernacular of the WiMAX Forum. Therefore, the WiMAX Forum has devised system profiles that specify combinations of parameters, such as operating frequency, modulation scheme and channelization. Presently, the WiMAX Forum has identified five profiles for 802.16-2004 that allow the technology to accommodate different frequency bands, channel bandwidths and duplexing schemes (TDD/FDD).

Interestingly, the 20 MHz radio channel that was required to achieve 70 Mbps of throughput is not one of the focus points at the moment. Some equipment providers are also currently targeting a 700 MHz solution for use in rural deployments, although it remains to be seen when, or even if, a profile is developed for this spectrum. (Note: 700 MHz is a very favourable spectrum for mobile use.)

When equipment companies unveil products built to these profiles, the forum will test them – in laboratories being set up now – for both conformance with the profiles and interoperability. The forum will credit products that pass with a 'WiMAX Certified' label.

The use of profiles is clearly needed in order to support a wide range of deployment options; in particular it reduces the abundance of options to a manageable number and also causes the industry to focus on those profiles that should be implemented first. WiMAX Forum has released profiles based on three spectrum bands for global deployment: 5, 3.5 and 2.5 GHz. Each of these bands is briefly discussed below.

Unlicensed 5 GHz

This frequency range includes bands between 5.25 and 5.85 GHz. In the upper 5 GHz band (5.725–5.850 GHz), many countries allow higher power output (4 W) which makes this band more attractive to WiMAX applications globally. Because much of this spectrum is unlicensed in the USA, it is readily available for WiMAX deployment without individual FCC licensing.

Licensed 3.5 GHz

Bands between 3.4 and 3.6 GHz have been allocated for BWA in the majority of countries, with the exception of the USA. In the USA, this spectrum is designated for use by the Federal Government. Because the WiMAX Forum is focusing its efforts internationally, however, much of this spectrum is available for WiMAX use in Europe and elsewhere.

Licensed 2.5 GHz

The bands between 2.5 and 2.69 GHz have been allocated in the USA, Mexico, Brazil and some Southeast Asian countries. In the USA, much of this spectrum has been licensed for use in the multipoint distribution service (MDS) and the instructional television fixed service (ITFS; Table 2.7).

Forum Certified™ Products

WiMAX Forum Certified™ products will be based upon a single global standard enabling complete interoperability worldwide. In an 802.16-based network, a network provider would set up base stations consisting of one or more sectors that are connected to their edge and core networks via wireless or wireline connections as appropriate. Each base station, capable of supporting hundreds of products vs. proprietary wireless access technologies, creates a strategic option for the network planner focused on developing a network fabric that is capable of scaling over time.

Products that have been through the WiMAX Forum Certified™ process will reduce investment uncertainties for all parties in the access network value chain, from technology providers to service providers to end users. For network operators, equipment and component

Table 2.7   Worldwide allocation of licensed and licence-exempt bands

Country/geographic area	Bands used (licensed and licence-exempt)
North America, Mexico	2.5 and 5.8 GHz
Central and South America	2.5, 3.5 and 5.8 GHz
Western and Eastern Europe	3.5 and 5.8 GHz
Middle East and Africa	3.5 and 5.8 GHz
Asian Pacific	3.5 and 5.8 GHz

manufacturers, and ultimately, for subscribers, products that are WiMAX Forum Certified[TM] will deliver wireless broadband access with a range of interoperable components.

Further, the evolution of a service provider's network over time being a key concern, especially the multi-vendor network that might result from standards-based products being available, will be solved by WiMAX Forum Certified[TM] products. For example, one of the building blocks of the IEEE 802.16 standard is the concept of a 'variable burst length,' a feature adopted to ensure a migration path from ATM networks to IP networks. With WiMAX Forum Certified[TM] products, service providers can be sure that this type of evolution can occur in their network even when it is made up of products from multiple vendors as certified interoperability will guarantee a known interoperability level between systems.

Emerging WiMAX Forum Certified[TM] products have significant advantages over current proprietary solutions:

- They create a forward looking transport and access network platform which is 'future-proof'.
- They provide a high speed internet access solution for the delivery of 'triple play' services (voice, video and data).
- They are generally based upon Ethernet, which has proven to be the most effective infrastructure for data networks over the past decade.
- They have the advantage of worldwide support from not only the WiMAX Forum[TM] but also the IEEE and ETSI.

WiMAX Forum Certified[TM] products are an enabler for a new generation of cooperative and strategic partnerships, bringing together service providers, network operators and equipment manufacturers to deliver service packages unrivalled by any other past offering in the wireless industry (Table 2.8).

What is Ahead?

IEEE 802.16 products are in the final phases of commercial development with initial trial deployments already underway. While the first wave of WiMAX certified products has been delayed due to delays in tests for certification and is anticipated during the early 2006, these tests will answer some of the strong critics of WiMAX by proving that not only

Table 2.8 Bands and Frequencies Available for WiMAX

Band	Frequencies	Licence required?	Availability
2.5 GHz	2.5–2.69 GHz	Yes	Allocated in Brazil, Mexico, some Southeast Asian countries and the U.S. (The WiMAX Forum also includes 2.3 GHz in this band category because it 'expects to cover [2.3 GHz] with the 2.5 GHz radio.') Ownership varies by country.
3.5 GHz	3.3–3.8 GHz, but primarily 3.4–3.6 GHz	Yes, in some countries	In most countries, the 3.4–3.6 GHz band is allocated for broadband wireless.
5 GHz	5.25–5.85 GHz	No	In the 5.725–5.85 GHz portion, many countries allow higher power output (4 W), which can improve coverage

does the wireless broadband technology work, but that the products coming out will be interoperable.

The forum expects the laboratory tests for certification in Malaga, Spain to be over soon. Then, the believers hope, some magic will occur. As soon as the 802.11 standard was established and the WiFi Alliance started certifying interoperability, an amazing phenomenon was witnessed in volume, price and adoption. The same dynamic will also be at work in case of WiMAX.

The forum has also started positioning WiMAX as 'personal broadband' to try to differentiate itself in the marketplace of the future, which implies a need for mobility – something WiMAX will not be able to deliver until 2007 or 2008 – but mobility is the direction in which many Forum members want to go. Intel, one of the prime movers behind the technology, expects to begin sampling WiMAX mobile chips in early 2006 and have it in laptops by 2007. WiMAX will be akin cell phones but for data (Figure 2.13).

2.8 WiMAX: REGULATION

The economic success of WiMAX may be impacted by the availability of licensed spectrum in some regions, although the impact of the availability and suitability of the spectrum resources on the success of WiMAX would be less in comparison to other wireless technologies.

Asia/Australia/New Zealand
- 1900–1920/2010–2025 MHz allocated to UMTS operator
- 2.5–2.7 GHz generally available for wireless broadband
- 3.4–3.6 GHz also available

Japan
- 1900–1920/2010–2025 MHz to be allocated soon

China
- 2.5–2.7 GHz owned by cable operators
- 3.4–3.6 GHz owned by major operators

Africa/Middle East
- 2.5–2.7 GHz MMDS auctions have been held in most countries
- 3.4–3.6 GHz also available in most countries

North America
- 2.5–2.7 GHz owned by Sprint, Nextel, and a few others in the US
- 3.4–3.6 GHz to be auctioned in Canada (uncertain future in the US)

Europe
- 1.9–1.92/2.01–2.025 GHz allocated to UMTS operators
- 2.5–2.7 GHz planned for UMTS extension in 2007
- 3.4–3.6 GHz held by insurgents or incumbent LECs in most countries

Latin America
- 2.5–2.7 GHz largely owned by insurgents
- 3.4–3.6 GHz mostly owned by fixed line incumbents

Figure 2.13 Global spectrum allocation

This is because spectrum options for WiMAX technology are extensive due to its broad operating range. However, as with any other spectrum-based technology, successful WiMAX deployment will depend upon spectrum resources to some extent.

Despite its wide range of spectrum options, the WiMAX Forum is focusing its efforts on the frequency range from 2 to 11 GHz.

Benefits of 2–11 GHz

At these frequencies, radio waves can penetrate some way into buildings, and can bend and reflect around obstacles to some extent, so the base station and client antennas do not need a clear line-of-sight between them, which is much more practical in an urban environment. However, there is also another significant consequence due to the nature of spectrum regulation.

Different versions of WiMAX are defined at these lower frequencies, partly as a result of differing national regulations governing wireless spectrum. For example, a frequency at 3.5 GHz requires a wireless operating licence and will have the full 50 km range (although most uses will probably be around 6–10 km). However, those at 2.4 or 5 GHz do not require a licence in most regulatory jurisdictions, because these frequency bands allow unlicensed use. In these bands, WiMAX operates at a much lower wireless power and has a very limited range (roughly the same as for WLAN, which also uses these frequency bands).

Deregulation

A major driver impacting the broadband wireless explosion is the advent of global telecom deregulation, opening up the telecommunications/Internet access industries to a host of new players. As more and more countries enable carriers and service providers to operate in a variety of frequencies, new and lucrative broadband access markets are springing up everywhere.

Wireless technology requires the use of frequencies contained within a given spectrum to transfer voice and data. Governments allocate a specific range of that spectrum to incumbent and competitive carriers, as well as cellular operators, ISPs and other service providers, enabling them to launch a variety of broadband initiatives based exclusively on wireless networking solutions. There are two main types of spectrum allocation: licensed and unlicensed.

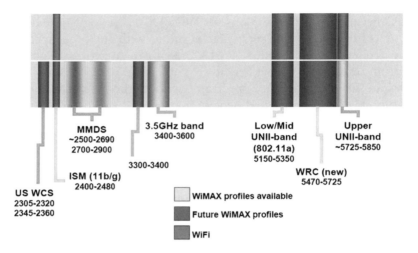

Figure 2.14   WiMAX spectrum

- Licensed frequencies are typically awarded through an auction or contest to those who present the soundest business plans to the regulatory authorities overseeing the process.
- Unlicensed frequencies allow multiple service providers to utilize the same section of the spectrum and compete with each other for customers.

Recent examples of the global spread of bandwidth allocations/ licences that are available to wireless operators as a result of deregulation include: Italy – 26 and 28 GHz bands; UK – 2.4, 3.5, 10.5 and 28 GHz bands; France – 2.4, 3.5 and 26 GHz bands; Sweden – 3.5 GHz band; EC – 5.4 GHz, to be made available for carriers throughout continental Europe; China – 2.4, 3.5, 5.8 and 26 GHz bands; and Brazil – 3.5 and 10.5 GHz bands (Figure 2.14).

Spectrum: Licensed and Unlicensed

Licensed spectrum requires an authorization/license from the regulators, which offers that individual user – or 'licensee' – the exclusive rights to operate on a specific frequency (or frequencies) at a particular location or within a defined geographic area. In contrast, unlicensed spectrum permits any user to access specific frequencies within a specified geographic area without prior regulatory authorization.

While users of this spectrum do not have to apply for individual licences or pay to use the spectrum, they are still subject to certain rules.

Table 2.9   List of first-stage system profiles used for WiMAX certification

Configuration			Profile name
3.5 GHz	TDD	7 MHz	3.5T1
3.5 GHz	TDD	3.5 MHz	3.5T2
3.5 GHz	FDD	3.5 MHz	3.5F1
3.5 GHz	FDD	7 MHz	3.5F2
5.8 GHz	TDD	10 MHz	5.8T

First, unlicensed users must not cause interference to licensed users and must accept any interference they receive. Second, any equipment that will be utilized on unlicensed spectrum must be approved in advance by the regulators (Table 2.9).

2.9   WiMAX ROLL-OUT

WiMAX Forum anticipates roll-out of its technology in three phases.

Phase I (Present–2005)

Fixed Location, Private Line Services, Hotspot Backhaul

Using the initial 802.16 standard as its cornerstone, phase I of WiMAX deployment has already begun with the provision of traditional dedicated-line services to carriers and enterprises. Phase I also includes such operations as aggregating public WiFi hotspots to a central, high-capacity Internet connection.

Phase II (2005–2006)

Broadband Wireless Access/Wireless DSL

Phase II of the roll-out will entail the first mass-market application of WiMAX technology. With the backing of computer industry heavyweights such as Intel Corporation and Dell, this phase will involve the delivery of low cost, user installable premises equipment that will not have to be pointed at a base station. In conjunction with the equipment roll-out, the Forum anticipates that the number of wireless internet service providers (WISPs) utilizing WiMAX compatible technology will increase exponentially.

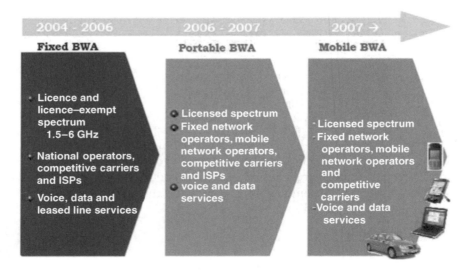

Figure 2.15   WiMAX roll-out phases

Phase III (2007)

Mobile/Nomadic Users

Phase III of the rollout will focus on the development of a mobile-broadband market. In this final phase, laptops and other mobile computing devices will be fully integrated with WiMAX chips and antennas, allowing mobile workers to send and receive high-bandwidth files such as schematics, videos and multimedia presentations in real time over a wireless broadband connection.

The WiMAX Forum anticipates that the technology will be deployed for the offering of other products and services, as well. For example, some believe that cellular operators will have the opportunity to decrease their independence on backhaul facilities leased from their competitors through the use of WiMAX technology. Still others believe that the technology will provide a secondary communications network for law enforcement, fire protection and other public safety organizations in congested metropolitan areas (Figure 2.15).

3

WiMAX Hype

WiMAX is coming, and along with it a great deal of vendor hype and operator confusion. WiMAX is the latest, and most-discussed, generation of wireless technology in years. Over the past few years there has been an extraordinary amount of hype and confusion surrounding WiMAX.

Unfortunately, this hype and confusion has resulted in the bar of hope being raised to such a high level that it will be very difficult for WiMAX to live up to expectations, despite potentially achieving a great deal. Further, because expectations have been raised to such a high level, there is a tendency to compare WiMAX with any and every wireless technology, irrespective of its area of application.

WiFi, MBWA, LMDS, Digital TV, 3G-based cellular and many present, future and past technologies are compared with WiMAX. Most of the time the conclusion reached is that WiMAX will compete with them or even replace them. WiMAX is a credible solution to a number of problems that have plagued the wireless industry since its inception. However, to judge it as a technology expected to make all other technologies vanish is taking things too far.

Without doubt WiMAX is capable of leading widespread adoption of broadband Internet access and changing the way PCs, handhelds and the World Wide Web are used but it will require more than one technology to alter how computer users live, work and play.

What is interesting is that the hype is based on future offerings expected from different flavours of WiMAX over the next couple of

The Business of WiMAX   Deepak Pareek
© 2006 John Wiley & Sons, Ltd

Table 3.1   BWA past failures and current drivers

Past reasons for failure	Current drivers of success
Overly aggressive projections for bandwidth growth	Rising bandwidth demand
Proprietary, high-cost equipment	Increasing need for connectivity
Inadequate equipment	Emerging standards
Poor implementations	Support from industry heavyweights Intel, Fujitsu, Alcatel and Siemens
High-spectrum costs	Capital availability

years. While initial signals coming from the field are positive, it is still too early to declare this technology a success.

WiMAX comprised a fixed wireless solution (802.16-2004) and a portable/mobile solution (802.16e). Given that there are more differences than similarities between the two solutions, it is only natural that some confusion exists. However, this confusion has also resulted in raised expectations that will be difficult, if not impossible, for WiMAX to achieve (Table 3.1).

3.1   THE CONFUSION(S)

The key confusion is regarding actual performance, with respect to distance as well as throughput. For example, WiMAX was originally billed as a wireless technology that could deliver 70 Mbps and extend coverage to 50 km, or roughly 30 miles. Most press reports also assumed that 70 Mbps would be achievable everywhere, including at the cell edge. In large part, little has been done to correct these misconceptions.

In order to achieve this level of performance, a fixed wireless point to point technology with LOS (line-of-sight) locations and directional antennas is required, meaning that all of the power is essentially dedicated to supporting that one connection – a rather expensive and impractical application for WiMAX under most scenarios.

Wireless backhaul can have notable exception where the subscription cost of the service offering could justify the dedicated resources. Achieving 70 Mbps in a mobile environment with WiMAX will not be feasible or economical in the foreseeable future.

Additionally, there is still an inherent tradeoff between data rates and distance, with the higher data rates only achievable near the centre of the cell. In other words, to achieve 70 Mbps data rates throughout an entire cell would require very small cell radii.

This level of performance is not extraordinary. There are also several non-WiMAX solutions, in particular microwave radio solutions, which can transmit a point-to-point signal that supports hundreds of Mbps or more. Simply crank up the power and allocate spectrum and virtually anything is possible. These high data rates may well be limited to fixed scenarios under certain special conditions, similar to those provided above. It is much more difficult to achieve high data rates in mobile environments.

3.2 THE 'AHAA(S)'

WiMAX is designed to cover wide geographical areas serving large number of users at low cost and provides a wireless alternative to wired backhaul and last-mile deployments that use data over cable service interface specification (DOCSIS) cable modems, digital subscriber line technologies (xDSL), T-carrier and E-carrier (T-x/E-x) systems, and optical carrier level (OC-x) technologies. WiMAX is considered one of the best solutions for last-mile distribution due to its extraordinary performance characteristics (Figure 3.1).

Throughput and Coverage

WiMAX technology can reach a theoretical 30 mile coverage radius and achieve data rates up to 75 Mbps, although at extremely long range

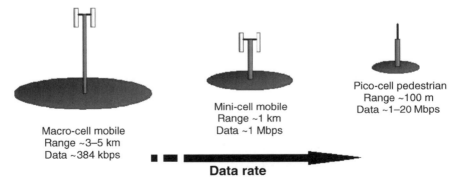

Figure 3.1 Coverage vs throughput

throughput is closer to the 1.5 Mbps performance of typical broadband services (equivalent to a T1 line). Dynamic adaptive modulation allows the base station to trade off throughput for range so that service providers can provision rates based on a tiered pricing approach, similar to that used for wired broadband services.

WiMAX 802.16 equipment certified by the Forum supports shared throughput of up to 75 Mbps and a coverage radius of 5–8 km (licence-exempt), and depends on terrain and population density. By using a robust modulation scheme, IEEE 802.16 delivers high throughput at long ranges with a high level of spectral efficiency that is also tolerant of signal reflections.

In addition to supporting a robust and dynamic modulation scheme, the IEEE 802.16 standard also supports technologies that increase coverage, including mesh topology and 'smart antenna' techniques. As radio technology improves and costs drop, the ability to increase coverage and throughput by using multiple antennas to create 'transmit' and/or 'receive diversity' will greatly enhance coverage in extreme environments.

Flexibility and Scalability

The 802.16-2004 standard supports flexible radio frequency (RF) channel bandwidths and reuse of these frequency channels as a way to increase network capacity. The standard also specifies support for transmit power control (TPC) and channel quality measurements as additional tools to support efficient spectrum use.

Easy addition of new sectors supported with flexible channels maximizes cell capacity, allowing operators to scale the network as the customer base grows. Flexible channel bandwidths accommodate spectrum allocations for both licensed and unlicensed spectrum.

The standard has been designed to scale up to hundreds or even thousands of users within one RF channel. Operators can reallocate spectrum through sectoring as the number of subscribers grows. Support for multiple channels enables equipment makers to provide a means addressing the range of spectrum use and allocation regulations faced by operators in diverse international markets.

The 802.16 standard provides an important flexibility advantage to new businesses or businesses that move their operations frequently, like a construction company with offices at each building site. Unlike a T1 or

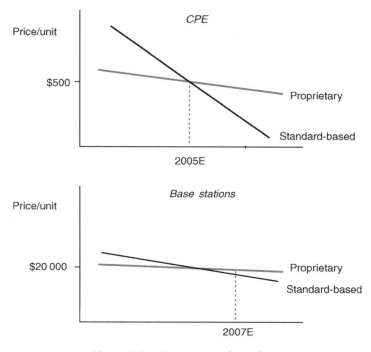

Figure 3.2 Coverage vs throughput

DSL line, wireless broadband access can be quickly and easily set up at new and temporary sites (Figure 3.2).

Cost Effectiveness

The wireless medium used by WiMAX enables service providers to circumvent costs associated with deploying wires, such as time and labour. Interoperable equipment allows operators to purchase WiMAX Certified™ equipment from more than one vendor. A stable, standard-based platform improves OpEx by sparking innovation at every layer, network management, antennas and more.

3.3 THE 'WHY(S)'

WiMAX is an integrated suite of many innovative and advance techniques covering diverse areas like modulation, antenna diversity and interference. Some of the key developments having direct bearing on commercial acceptance of system functionality are as follows.

Dynamic Burst Mode TDMA MAC

802.16 is optimized to deliver high, bursty data rates to subscriber stations. This means that IEEE 802.16 is uniquely positioned to extend broadband wireless beyond the limits of today's systems, both in distance and in the ability to support applications.

Quality of Service

Voice capability is extremely important, especially in underserved international markets. For this reason the IEEE 802.16a standard includes quality of service features that enable services including voice and video that require a low-latency network.

The grant/request characteristics of the 802.16 MAC enable an operator to simultaneously provide premium guaranteed levels of service to businesses, such as T1-level service, and high-volume 'best effort' service to homes, similar to cable-level service, all within the same base station service area cell.

Link Adaptation

Adaptive modulation and coding occur subscriber by subscriber, burst by burst, uplink and downlink. Transmission adaptation, with the help of modulation depending on channel condition, provides high reliability to the system. It also keeps more users connected by virtue of its flexible channel widths and adaptive modulation. Because it uses channels narrower than the fixed 20 MHz channels used in 802.11, the 802.16-2004 standards can serve lower data-rate subscribers without wasting bandwidth. When subscribers encounter noisy conditions or low signal strength, the adaptive modulation scheme keeps them connected when they might otherwise be dropped.

Further, this feature imparts differential service provision to the system, making it economically more appealing to operators. Dynamic adaptive modulation allows the base station to trade off throughput for range. For example, if the base station cannot establish a robust link to a distant subscriber using the highest order modulation scheme, 64 QAM, the modulation order is reduced to 16 QAM or QPSK (quadrature phase shift keying), which reduces throughput and increases effective range.

NLOS Support, Provides Wider Market and Lower Costs

The problems resulting from NLOS conditions are solved or mitigated by using multiple frequency allocation support from 2 to 11 GHz, OFDM and OFDMA for NLOS applications (licensed and licence-exempt spectrum), subchannelization, directional antennas, transmit and receive diversity, adaptive modulation, error correction techniques and power control (Figure 3.3).

Highly Efficient Spectrum Utilization, Provides High Efficiency

MAC, designed for efficient use of spectrum, incorporates techniques for efficient frequency reuse, deriving more efficient spectrum usage of the access system.

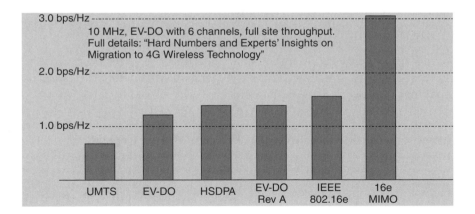

Figure 3.3 Spectrum efficiency

Flexible Channel Bandwidth, Provides Bandwidth on Demand and Scalability

As the distance between a subscriber and the base station (or AP) increases, or as the subscriber starts to move by walking or driving in a car, it becomes more of a challenge for that subscriber to transmit successfully back to the base station at a given power level. For power-sensitive platforms such as laptop computers or handheld devices, it is often not possible for them to transmit to the base station over long distances if the channel bandwidth is wide.

The IEEE 802.16-2004 and IEEE 802.16e standards have flexible channel bandwidths between 1.5 and 20 MHz to facilitate transmission over longer ranges and to different types of subscriber platforms. In addition, this flexibility of channel bandwidth is also crucial for cell planning, especially in the licensed spectrum. For scalability, an operator with 14 MHz of available spectrum, for example, may divide that spectrum into four sectors of 3.5 MHz to have multiple sectors (transmit/receive pairs) on the same base station. With a dedicated antenna, each sector has the potential to reach users with more throughput over longer ranges than an omnidirectional antenna can. Net-to-net, flexible channel bandwidth is imperative for cell planning.

Smart Antenna Support, Provides Better Throughput-to-range Relationship

Smart antennas are being used to increase the spectral density (i.e. the number of bits that can be communicated over a given channel in a given time) and the signal-to-noise ratio for both Wi-Fi and WiMAX solutions. Because of performance and technology, the 802.16-2004 standard supports several adaptive smart antenna types, including receive spatial diversity antennas, simple diversity antennas, beam-steering antennas and beam-forming antennas.

Error Correction Techniques, Provide Better Overall Performance

Error correction techniques have been incorporated into WiMAX to reduce the system signal-to-noise ratio requirements. Forward error correction, convolutional encoding and interleaving algorithms are used to detect and correct errors to improve throughput. These robust error correction techniques help to recover erroneous frames that may have been lost due to frequency selective fading or burst errors. Automatic repeat request (ARQ) is used to correct errors that cannot be corrected by the FEC, by having the erroneous information resent. This significantly improves the bit error rate (BER) performance for a similar threshold level.

Power Control, Provides Better Power Efficiency and Coexistence

Power control algorithms are used to improve the overall performance of the system; it is implemented by the base station sending power control information to each of the CPEs to regulate the transmit power level so that the level received at the base station is at a predetermined level. In a dynamical changing fading environment, this predetermined performance level means that the CPE only transmits enough power to meet this requirement. The converse would be for the CPE transmit level to be based on worst-case conditions. The power control reduces the overall power consumption of the CPE and the potential interference with other colocated base stations. For LOS the transmit power of the CPE is approximately proportional to its distance from the base station; for NLOS it is also heavily dependent on the clearance and obstructions.

Data Security, Provides Secure Communication

WiMAX proposes a full range of security features to ensure secured data exchange: terminal authentication by exchanging certificates to prevent rogue devices, user authentication using the EAP, data encryption using the data encryption standard (DES) or advanced encryption standard (AES), both much more robust than the WEP initially used by WLAN. Furthermore, each service is encrypted with its own security association and private keys.

Since its early days, IEEE 802.16 standards have seen many changes. Even today they continue to innovate and evolve with time and technology advancements. Security concern for wide-range connectivity is always there. The security features of the standard are being worked on so that, when its time for the product release to market, the technology will be more secure.

Backing from Giants Industry, Provides Better Value for Money

Intel is actively participating in WiMAX industry efforts to help reduce investment risks for operators and service providers while enabling them to more cost effectively take advantage of the tremendous market potential of wireless broadband access. The 802.16 wireless standard will provide a flexible, cost-effective means of filling existing gaps in broadband coverage, and creating new forms of broadband services, not thought of in a wired world. With Intel, Fujitsu and Nokia backing this wireless technology and standards, 802.16 and its variants will find many takers in the product arena.

3.4   THE 'OOP(S)'

Some of the challenges faced by WiMAX are as follows.

RF Interference

An interfering RF source disrupts a transmission and decreases performance by making it difficult for a receiving station to interpret a signal. Forms of RF interference frequently encountered are multipath interference and attenuation. Multipath interference is caused by signals reflected from objects, resulting in reception distortion. Attenuation occurs when an RF signal passes through a solid object, such as a tree, reducing the strength of the signal and subsequently its range. Overlapping interference from an adjacent base station can generate random noise.

Licence-exempt solutions have to contend with more interference than licensed solutions, including intranetwork interference caused by the service provider's own equipment operating in close proximity and external network interference. Licensed solutions must only contend with internetwork interference. For licence-exempt solutions, RF interference is a more serious issue in networks with centralized control than in a shared network because the base station coordinates all traffic and bandwidth allocation.

Addressing issues with interference

Interference is the disruption or degradation of a transmitted signal by extraneous RF energy. Interference impedes the ability of an RF receiver to distinguish between the transmitted signal and the background RF energy that exists at that specific point in time. Causes of extraneous RF energy include noise, and direct spectrum overlap by identified and unidentified sources.

Extraneous RF energy can be addressed by subchannelization and adaptive modulation, proper network design, filtering, shielding, synchronization of signals and the use of power amplifiers and antenna technologies.

Infrastructure Placement

Infrastructure location refers to the physical location of infrastructure elements. Infrastructure placement can be an issue for both licensed and licence-exempt solutions. However, infrastructure placement presents some special considerations for licence-exempt solutions. Service providers are quickly deploying solutions in specific areas to stake out territories with high subscriber density and spectrum efficiency. Such areas include higher ground, densely populated or population growth areas, and areas with a less crowded RF spectrum. In addition, the physical structure that houses or supports the base station must be RF-compatible. A metal farm silo, for example, may distort signals, or a tree swaying in the wind may change signal strength.

Addressing issues with infrastructure placement

Infrastructure placement establishes the foundation for the service provider's network. When choosing a location for deployment, a service provider must ensure that it can obtain access to the site at all times, that the building or location does not contain physical material that is not RF-friendly and that the infrastructure provides protection against weather-related elements, such as wind and lightning. Obstacles such as trees and buildings frequently block signal paths in urban areas and some rural areas. NLOS performance is greatly improved with 802.16-2004 due to its improved resistance to multipath interference. Even with no direct LOS between the base station and the subscriber station, signals can be received after they reflect off buildings or other obstructions. Factors such as these make a preliminary site survey indispensable. Infrastructure placement provides a solid market advantage for incumbents. The cost and time involved in obtaining building permits, leases and roof space present significant barriers to those without an already established infrastructure.

Deployment Costs

The increased number of cell sites, as a result of using higher frequency bands, increases site acquisition/leasing and construction costs, regardless of the technology being deployed. The cost to acquire a site in North America can easily reach $25 000, plus ongoing lease costs, while an operator may have to spend up to $75 000 on construction costs to get the site up and running – assuming the operator starts from scratch. Further, the logistical challenges of getting enough sites to deploy a ubiquitous mobile network can pose a tremendous challenge, regardless of the cost factor.

That said, WiMAX will probably have a lower cost structure with respect to the core network, or the portion of the network that is 'behind' the base stations. Specifically, WiMAX uses an all-IP core which means it is scalable and can therefore support a higher level of user traffic for a given amount of network resources.

Additionally, WiMAX makes use of off-the-shelf routers vs a combination of circuit switches and other network components, which, although similar to off-the-shelf routers, have been specially customized for use in a cellular network. It is important to point out, however, that 3G is also transitioning to an all-IP core, at which point it will greatly reduce its own cost structure and achieve higher scalability than is possible today.

Unfinished Business

The 802.16e standard only addresses the PHY and MAC layers, leaving it to the WiMAX Forum to tackle issues such as call control, session management, security, the network architecture and roaming. To put things in perspective, as the standard is currently written, each WiMAX base station is virtually oblivious of its surrounding base stations while the MAC layer only has placeholders for the messaging traffic associated with implementing a handover. As a consequence, the notion of seamless mobility does not exist while power management issues could result in reduced performance, in particular for users at the cell edge (25–35% of the network), where intercell interference would be most evident.

The WiMAX Forum created a network architecture working group in late 2004 to address some of these unresolved issues, but it is unrealistic to expect all of them to be solved, let alone tested and verified, in a few months. As it stands now, the first revision of the networking specification is scheduled to be completed by the end of the year. As an interim step, the WiMAX Forum is moving to first implement a portable solution which lacks some of the network intelligence required to support higher vehicular speeds (up to 120 km/h) and seamless handoffs.

In lieu of applications and services, such as voice, that require seamless handoffs and in the absence of widespread coverage, a portable broadband connection should more than adequately meet the needs of high-bandwidth data users.

WiMAX Chipset Availability

Another major uncertainty is the availability of chipsets. In addition to Intel and Fujitsu, several private companies are also promising very compelling .16e chipset solutions and they may, in fact, beat the larger silicon suppliers to the market. Regardless of who is first to market, it will be a challenge to have the silicon available for sampling any time soon. The mobile standard will not be finished until later in 2005 and the initial profiles have not been selected yet, meaning that, while some work can currently be done, the fine technical details cannot be implemented until after the standard is fully ratified. Equally important, the major semiconductor companies who are important to the success of WiMAX are not necessarily first-to-market suppliers of wireless chipsets (Wi-Fi and cellular technologies are two examples). Given some of the requirements for the mobile WiMAX solution, it could take more than one die spin to manufacture a chipset that supports the initial WiMAX profiles and does so with adequate performance (size, power requirements etc.).

Interoperability Testing and Market Feel

While the 802.16e standard could be completed in late 2005, it does not necessarily suggest that the technology will then be ready for commercial deployment. Even for the 802.16d standard, multivendor interoperability tests, commonly referred to as 'Plugfests', are yet to occur, although they are expected to begin later this year.

Interoperability testing always takes longer than anticipated, in particular if an entirely new standard is being tested and if companies not normally accustomed to this type of activity are involved. Assuming that interoperability testing is successful and that commercially viable solutions (e.g. data cards) are available, potential operators could then take months conducting field trials before moving to a market trial and then potentially a wider-scale commercial roll-out.

It is interesting to note that current plans for WiMAX plugfests are to certify equipment against one of the many WiMAX targeted profiles. Since the Forum targets multiple profiles for different regions and applications, many interoperability activities will be required. Additionally, end-to-end plugfests cannot be a reality until WiMAX base stations and WiMAX CPEs are available. If history is a guide, the WiMAX base stations will be ready for interoperability testing well before the CPEs will be ready.

Uncertain Economics

Like with other wireless technologies, the economics of using WiMAX to offer fixed wireless services in regions of the world where wireline deployments have not taken place or where there is little competition are attractive. By eliminating the need to deploy copper or fibre, an operator can significantly reduce its upfront capital expenditures while at the same time reduce the risk of service disruption due to vandalism or theft of the buried cabling.

Once consumers self-install the CPE, the deployment cost advantages become even more compelling. It is not clear if the same can be said for other market opportunities, especially when the network operator is designing its network to support seamless mobility and voice – far more base stations are required, regardless of the air interface that is used. However, if the operator deploys its WiMAX network in selected, albeit geographically large, areas where portable/mobile broadband data traffic is highest, and if the operator does not attempt to deliver ubiquitous coverage within that area, its cost structure will be reduced.

Put simply, deploying a mobile network is not an inexpensive proposition, and with an abundance of mobile operators in most countries, these regions may not be able to support another greenfield mobile operator. These regions could, however, support a service that differentiated itself by offering higher data rates, with the tradeoff coming in the form of reduced coverage and lower quality of service – seamless handoffs, high-speed vehicular support, etc.

PART Two
WiMAX Effect

4

WiMAX Solutions

IEEE 802.16 is an emerging suite of air interface standards for combined fixed, portable and MBWA. Initially conceived as a radio standard to enable cost-effective last-mile broadband connectivity to those not served by wired broadband such as cable or DSL, the specifications are evolving to target a broader market opportunity for mobile, high-speed broadband applications. The realization of a low-cost, broadly interoperable wide-area data network that supports portable and mobile usage could have significant end-user benefits.

Imagine a radio access network that provides broadband access to users at home, in the office, in areas underserved by wireline services and even to users on the move equipped with portable devices like laptops, PDAs and smart phones. WiMAX can provide a flexible radio access solution that offers these features, based on an attractive full IP architecture delivering the capacity required to support wireless broadband services.

WiMAX is at the centre of the emergence of new market and technology opportunities. The widespread deployment of high-speed Internet at home has opened the door to the introduction of new services, such as video, audio, gaming and e-commerce. Today, the availability of portable devices like laptops, PDAs and smart phones is generating interest in providing similar services under nomadic conditions.

After the emergence and wide acceptance by users of Internet, broadband and mobile services, we can anticipate a future need for nomadic broadband wireless services. By addressing multiple market segments through the standardization and interoperability efforts of the WiMAX

The Business of WiMAX   Deepak Pareek
© 2006 John Wiley & Sons, Ltd

Forum, volume production of WiMAX Certified equipment will become possible, driving down the total cost of ownership and opening up new opportunities for the delivery of broadband services to bridge the digital divide.

WiMAX access can be easily integrated within both fixed and mobile architectures, enabling operators to integrate it within a single converged core network, thereby providing new capabilities for a user-centric broadband world.

Initial deployments of IEEE 802.16 standard-based networks will probably target fixed access connectivity to unserved and underserved markets where wireline broadband services are insufficient to fulfil the market need for high-bandwidth Internet connectivity.

Precertification WiMAX and prestandards WiMAX-type proprietary implementations exist today, which are addressing this fixed access service environment. On completion, standardization and certification will help accelerate the ramp for these fixed access solutions by providing interoperability amongst equipment and economies of scale resulting from high-volume standard-based components.

While many technologies currently available for fixed broadband wireless can only provide LOS coverage, the technology behind WiMAX has been optimized to provide excellent NLOS coverage. WiMAX's advanced technology provides the best of both worlds – large coverage distances of up to 50 km under LOS conditions and typical cell radii of up to 5 miles/8 km under NLOS conditions (Figure 4.1).

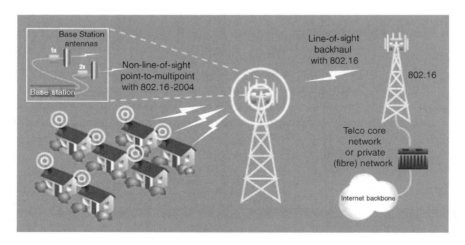

Figure 4.1   LOS and NLOS installation

4.1 LOS

The radio channel of a wireless communication system is often described as being either LOS or NLOS. In an LOS link, a signal travels over a direct and unobstructed path from the transmitter to the receiver.

In an NLOS link, a signal reaches the receiver through reflections, scattering and diffractions. The signals arriving at the receiver consist of components from the direct path, multiple reflected paths, scattered energy and diffracted propagation paths. These signals have different delay spreads, attenuation, polarizations and stability relative to the direct path.

The multipath phenomena can also cause the polarization of the signal to be changed. Thus, using polarization as a means of frequency reuse, as is normally done in LOS deployments, can be problematic in NLOS applications. How a radio system uses these multipath signals to advantage is the key to providing service under NLOS conditions.

A product that merely increases power to penetrate obstructions (sometimes called 'near LOS') is not NLOS technology because this approach still relies on a strong direct path without using energy present in the indirect signals. Both LOS and NLOS coverage conditions are governed by the propagation characteristics of their environment, path loss and radio link budget.

WiMAX technology can provide coverage under both LOS and NLOS conditions. NLOS has many implementation advantages that enable operators to deliver broadband data to a wide range of customers.

4.2 BENEFITS OF NLOS

There are several advantages that make NLOS deployments desirable. For instance, strict planning requirements and antenna height restrictions often do not allow the antenna to be positioned for LOS. For large-scale contiguous cellular deployments, where frequency reuse is critical, lowering the antenna is advantageous to reduce the cochannel interference between adjacent cell sites. This often forces the base stations to operate under NLOS conditions. LOS systems cannot reduce antenna heights because doing so would impact the required direct view path from the CPE to the base station.

NLOS technology also reduces installation expenses by making under-the-eaves CPE installation a reality and easing the difficulty of locating adequate CPE mounting locations. The technology also reduces the need for preinstallation site surveys and improves the accuracy of NLOS planning tools.

The NLOS technology and the enhanced features in WiMAX make it possible to use indoor CPE. This has two main challenges: firstly, overcoming the building penetration losses and secondly, covering reasonable distances with the lower transmit powers and antenna gains that are usually associated with indoor CPEs. WiMAX makes this possible, and the NLOS coverage can be further improved by leveraging some of WiMAX's optional capabilities.

WiMAX technology has many advantages that allow it to provide NLOS solutions with essential features such as OFDM technology, adaptive modulation and error correction. Furthermore, WiMAX has many optional features, such as ARQ, subchannelling, diversity and space–time coding that will prove invaluable to operators wishing to provide quality and performance that rivals wireline technology.

4.3   SELF-INSTALL CPE

Self-install CPEs are NLOS units which do not require specialized installation. The user can install such CPE with ease as no special skills are required. The biggest benefit of self-install CPE is that it is very cost-effective as no expenditure is incurred for installation and maintenance. Other benefits include user comfort, higher reliability and simplicity.

The challenge is that the industry will need to generate CPE – especially indoor, consumer-installable CPE – at viable prices. A real stumbling block in the residential space is the self-install issue. Without such user-friendly equipment, it would be quite difficult, not only for WiMAX, but for any technology to become a mass-market success. Yet a device that can work anywhere within a home presents a difficult engineering task. The challenge of meeting system-level requirements becomes larger as you try to penetrate walls and compensate for different materials in those walls. Terrain also plays a big role, as does the service provider's coverage in that area. All said, an outdoor antenna will always provide better reception.

Moving on to the question of portable clients, more challenges arise. The overall power budget you have to design for is obviously less than that for a device you are going to plug into the wall. Thermal and electrical envelopes become tighter too, and issues such as authentication and handoff from cell to cell and network to network will require attention.

In general, supporting indoor antennas and portable-system users requires smaller cells, which translates into more base stations and higher capital expenditures.

Again, these challenges are not insurmountable, but progress will take time. Vendors are confident that technologies such as MIMO (multiple-in, multiple-out) antennas will meet these challenges. Further, by using an on-chip system and investing in integration work, the cost can be reduced to an attractive level.

The wish list for WiMAX CPE is:

- lower cost;
- plug and play;
- greater throughput;
- quality of Service for value-added services;
- rapid shipping.

4.4   NOMAD, PORTABLE AND MOBILE DEPLOYMENT

As IEEE 802.16 solutions evolve to address portable and mobile applications, the required features and performance of the system will increase. Beyond fixed access service, even larger market opportunities exist for providing cost-effective broadband data services to mobile users. Initially this includes portable connectivity for customers who are not within reach of their existing fixed broadband or WLAN service options. This type of service is characterized by access that is unwired but stationary in most cases, albeit with some limited provisions for user mobility during the connection.

In this manner, 802.16 can be seen as augmenting coverage of 802.11 for private and public service networks and cost-effectively extending hotspot availability to wider ranges of coverage. Based on this described capability, this phase of deployment is referred to as 'portability with simple mobility'. The next phase of functionality, known as 'full mobility', provides incremental support for low-latency, low-packet-loss real-time handovers between access points at speeds of 120 km/h or higher, both within and between networks. This will deliver a rich end-user experience for high-quality multimedia applications.

The security mechanisms within the IEEE 802.16-2004 standard may be adequate for fixed access service, but need to be enhanced for

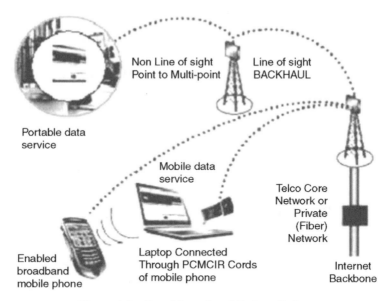

Figure 4.2   Portable and mobile installation

portable and mobile applications. An additional challenge for the fixed access air interface as well as subsequent portable and mobile services is the need to establish high-performance radio links capable of data rates comparable to wired broadband service, using equipment that can be self-installed indoors by users, as is the case for DSL and cable modems. Doing so requires advanced PHY layer techniques to achieve link margins capable of supporting high throughput in NLOS environments.

As 802.16 technologies evolve to address portable and mobile service, so do the feature requirements of the air interface and RAN network, interoperability demands and interworking with other dissimilar networks like Wi-Fi and 3G. The simple fact that mobile clients can dynamically associate and perform handover across APs, crossing large, possibly discontinuous geographic regions and operator domains, drives the need for a number of network-related enhancements (Figure 4.2).

4.5   NOMADICITY AND PORTABILITY

The simplest case of portable service (referred to as nomadicity) involves a user transporting an 802.16 modem to a different location. Provided this visited location is served by wireless broadband service, in this scenario, the user reauthenticates and manually reestablishes new IP connections and is afforded broadband service at the visited location. This usage enhancement over fixed access requires enhancements to security such as strong mutual authentication between the user/client device and the network AP supporting a flexible choice of credential types.

The next stage, portability with simple mobility, describes a more automated management of IP connections with session persistence or automatic reestablishment following transitions between APs. This incremental enhancement allows for more user transparent mobility and is suitable for latency tolerant applications such as TCP; it does not provide adequate handover performance for delay and packet-loss-sensitive real-time applications such as VoIP.

4.6 MOBILITY

In the fully mobile scenario, user expectations for connectivity are comparable to those experienced in 3G voice/data systems. Users may be moving while simultaneously engaging in a broadband data access or multimedia streaming session. The need to support low-latency and low-packet-loss handovers of data streams as users transition from one AP to another is clearly a challenging task.

For mobile data services, users will not easily adapt their service expectations because of environmental limitations that are technically challenging but not directly relevant to the user (such as being stationary or moving). For these reasons, the network and air interface must be designed upfront to anticipate these user expectations and deliver accordingly.

Of the three PHY layers supported in the standard, scalable OFDMA is the most versatile and the one preferred for high-speed mobile operation. Scalable OFDMA supports features (enhanced over OFDM) such as downlink (DL) and uplink (UL) subchannelization, fixed subcarrier spacing, and reduced overhead for cyclic prefix (CP) by keeping its duration constant at one-eighth the OFDMA symbol duration.

The 802.16 MAC is designed for PMP applications and is based on CSMA/CA. The 802.16 AP MAC manages UL and DL resources including transmit and receive scheduling. The MAC incorporates several features suitable for a broad range of applications at different mobility rates, such as the following:

- four service classes – UGS, rtPS, nrtPS and BE;
- header suppression, packing and fragmentation for efficient use of spectrum;
- PKM for MAC layer security; PKM version 2 incorporates support for EAP;
- broadcast and multicast support;
- high-speed handover and mobility management primitives;
- three power management levels: normal operation, sleep and idle (with paging support).

These features combined with the inherent benefits of scalable OFDMA make 802.16 suitable for high-speed data and bursty or isochronous IP multimedia applications.

4.7 SPECTRUM

Governments around the world have established frequency bands available for use by licenced and licence-exempt WiMAX technologies. Each geographical region defines and regulates its own set of licensed and licence-exempt bands. To meet global regulatory requirements and allow providers to use all available spectrums within these bands, the 802.16-2004 standard supports channel sizes between 1.5 and 20 MHz.

What is sometimes overlooked is that each band provides a different set of advantages for different usage models. Each serves a different market need based on tradeoffs between cost and QoS. Licence-exempt solutions and licensed solutions offer certain advantages to providers. The availability of both allows providers and emerging markets to fulfil a variety of usage needs.

4.8 LICENSED SPECTRUM

The 2.5 GHz band has been allocated in much of the world, including North America, Latin America, Western and Eastern Europe and parts of the Asia-Pacifi region as a licensed band for WiMAX. Each country allocates the band differently, so the spectrum allocated across regions can range from 2.6 to 4.2 GHz.

In the USA, the Federal Communications Commission (FCC) has created the Broadband Radio Service (BRS), previously called the multichannel multipoint distribution system (MMDS), for wireless broadband access. The restructuring that followed has allowed for the opening of the 2.495–2.690 GHz bands for licensed solutions such as 2.5 GHz in WiMAX. In Europe, ETSI has allotted the 3.5 GHz band, originally used for wireless local loop (WLL), for licensed WiMAX solutions.

To deploy a licensed solution, an operator or service provider must purchase spectrum. The purchasing of spectrum is a cumbersome process. In some countries, filing the appropriate permits to obtain licensing rights may take months, while in other countries, spectrum auctioning can drive up prices and cause spectrum acquisition delays.

Benefits of Licensed WiMAX Systems

A WiMAX system operating in the licensed band has an advantage over a system operating in an unlicensed band in that it has a more generous downlink power budget and can better support indoor antennas. Another significant advantage is the lower frequencies associated with licensed bands (2.5 and 3.5 GHz) enable better NLOS and RF penetration.

The higher costs and exclusive rights to spectrum enable a more predictable and stable solution for large metropolitan deployments and mobile usage. Higher barrier to entrance, coupled with exclusive ownership of a band, enables service quality improvements and reduces interference.

However, licensed bands are not without interference issues. As service providers deploy more networks, they must contend with mutual interference resulting from within their own network. Proper design and implementation can alleviate these problems. In summary, licensed solutions offer improved QoS advantages over licensed exempt solutions.

4.9   LICENCE-EXEMPT

The most commonly discussed unlicensed band, available virtually worldwide today, is in the vicinity of 2.4 GHz. This band is often called the industrial, scientific, medical (ISM) band because its initial allocation was to allow radio emissions by various sorts of equipment. This is the band that is being used today for WLAN according to the IEEE 802.11b/g standards and has been branded by an industry group as Wi-Fi.

Another commonly discussed set of bands is in the space between 5 and 6 GHz where the IEEE 802.11a standard is defined to operate. The unlicensed allocations in this band have been the subject of recent international harmonization efforts through the ITU at the 2003 World Radiocommunication Conference (WRC-03).

The majority of countries around the world have embraced the 5 GHz spectrum for licence-exempt communications. The 5.15 and 5.85 GHz bands have been designated as licence exempt in much of the world. Approximately 300 MHz of spectrum is available in many markets globally, and an additional 255 MHz of licence-exempt 5 GHz spectrum is available in highly populated markets like the USA.

Some governments and service providers are concerned that interference resulting from the availability of too many licence-exempt bands could affect critical public and government communication networks, such as radar systems. These countries and entities have become active in establishing limited control requirements for 5 GHz spectrums. For example, the UK is currently introducing restrictions on certain 5 GHz channels and considering enforcement of the use of the DFS function.

One key point which needs emphasis is that unlicensed does not mean unregulated, and all operators providing wireless services still need to maintain a no-interference working plan and a 'good neighbour' attitude along with efficient spectrum utilization.

Benefits of Licence-exempt WiMAX Systems

The costs associated with acquiring licensed bands are leading many WISPs and vertical markets to consider licence-exempt solutions for specialized markets, such as rural areas and emerging markets.

Licence-exempt solutions provide several key advantages over licensed solutions, including lower initial costs, faster roll-out and a common band that can be used in much of the world. These benefits are fuelling interest and have the potential to accelerate broadband adoption. Service providers in emerging markets, such as developing countries or mature countries with underdeveloped areas, can reduce time to market and initial costs by quickly deploying a licence-exempt solution without timely permits or auctions. Even mature areas can benefit from licence-exempt solutions.

Some service providers can use a licence-exempt solution to provide last-mile access for home, business or backhaul or as a supplemental network backup for their licensed or wired networks. A licence-exempt solution is regulated in terms of the transmission output power, although a permit is usually not required. A device or service can use the band at any time as long as output power is controlled adequately.

Providers who are particularly concerned about QoS, for example, may find that a licensed solution provides them with more control over the service. A service provider wanting to serve an emerging or under-developed market with a business class service can use a licence-exempt solution, with proper network design including site surveys and specialized antenna solutions, to offer certain Service Level Agreements (SLAs) for their specialized markets.

5

WiMAX Applications

The 802.16 standard will help the industry provide solutions across multiple broadband segments. WiMAX was developed to become a last-mile access technology comparable to DSL, cable and T1 technologies. It is a rapidly growing technology that is most viable for backhauling the rapidly increasing volumes of traffic being generated by Wi-Fi hotspots.

WiMAX is a MAN technology that fits between wireless LANs, such as 802.11, and wireless wide-area networks (WANs), such as the cellular networks. Bandwidth generally diminishes as range increases across these classes of networks. Proponents believe that WiMAX can serve in applications such as cellular backhaul systems, in which microwave technologies dominate, backhaul systems for Wi-Fi hot spots and most prominently as residential and business broadband services.

WiMAX is billed to support many types of wireless broadband connections including but not limited to the following: high-bandwidth MANs, cellular backhaul, clustered Wi-Fi hotspot backhaul, last-mile broadband, cell phone replacements and other miscellaneous applications such as automatic teller machines (ATMs), vehicular data and voice, security applications and wireless VoIP. Today, wherever available, these applications use expensive, proprietary methods for broadband access (Figure 5.1).

WiMAX was developed to provide low-cost, high-quality, flexible, BWA using certified, compatible and interoperable equipments from multiple vendors. As WiMAX is based on interoperability-tested systems that were built using the IEEE 802.16 standard-based silicon solutions, WiMAX will reduce costs. WiMAX is well placed to address

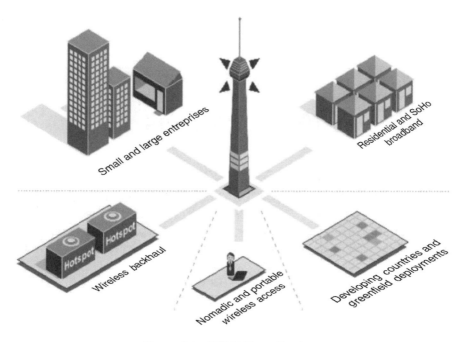

Figure 5.1   WiMAX applications

challenges associated with traditional wired access deployment types such as:

- large area coverage access, covering a large area (also referred to as hot zones) around the base station and providing access to 802.16 REV E clients using point-to-multipoint topology;
- last-mile access, connecting residential or business subscribers to the base station using point-to-multipoint topology;
- backhaul, connecting aggregate subscriber sites to each other and to base stations across long distances using point-to-point topology.

Let us understand these applications and how WiMAX provides a compelling business case for each of them.

5.1   METROPOLITAN-AREA NETWORKS

What makes WiMAX so attractive is its potential to provide broadband wireless access to entire sections of metropolitan areas as well as small

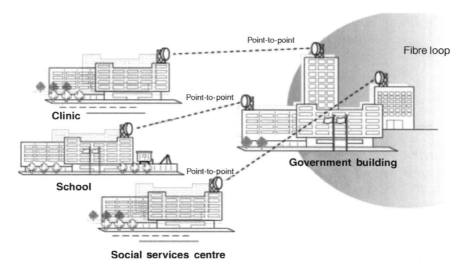

Figure 5.2 Metropolitan-area networks

and remote locales throughout the world. People who could not afford broadband will now be able to get it, and in places where it may not previously have been available. WiMAX enables coverage of a large area very quickly (Figure 5.2).

Today, MANs are being implemented by a wide variety of innovative techniques such as running fibre cables through subway tunnels or using broadband over power lines (BPL). In response to these new techniques, there has been a growing interest in the development of wireless technologies that achieve the same results as traditional MANs without the difficulty of supplying the actual physical medium for transmission, such as copper or fibre lines.

Undeniably, wireless MANs (WMANs) are emerging as a viable solution for broadband access. MANs are intended to serve an area approximately the size of a large city; MANs serve as the intermediary network between LANs and WANs. WMANs consist of a fixed wireless installation that interconnects locations within a large geographic region.

Despite the challenges, wireless metro-access solutions are continuously sought after as they are more cost-effective and flexible than their wired counterparts.

- WISPs can offer broadband services to geographically challenged areas (such as rural towns).

- Local governments can provide free access for businesses or emergency services (such as police and fire-fighters).
- Educational institutions can broaden learning through online collaboration between students and faculty on and off campus.
- Enterprises and large private networks can communicate and monitor supply-chain activities in near real time.

Wireless internet service providers (WISPs) have been striving for wireless technologies that make wireless metro access possible. Access to areas that are too remote, too difficult or too expensive to reach with traditional wired infrastructures (such as fibre) require new technologies and a different approach. The three key deployment types that make up wireless metro access are backhaul, last-mile and large-area coverage (referred to as hot zones).

Broadband wireless access provides more capacity at lower cost than DSL or cable for extending the fibre networks and supporting multimedia and fast Internet applications in the enterprise or home, but it has been held back by the lack of a standard, so that solutions have been based on proprietary, single-vendor efforts. Standardization through the IEEE 802.16 specification raises the potential to:

- make wireless the key platform of the future by providing more value than wired broadband;
- extend the range of Wi-Fi so that the dream of ubiquitous wireless can become a reality and provide an alternative or complement to 3G;
- provide an economically viable communications infrastructure for developing countries and mobile black spot regions in developed nations.

IEEE 802.16-based networks address the last mile of the communications infrastructure between a service provider's point of presence (POP) and business or residential customer locations. In residential areas today, the last mile, also known as the access network or local loop, consists predominantly of wireline choices – copper telephone wires or coaxial cable-TV cables. In metropolitan areas, where there is a high concentration of business customers, the wireline access network today often includes high-capacity SONET rings, optical T3 (45 mbps), and copper-based T1 (1.5 mbps) lines. Generally, these last-mile connections for small and medium-sized enterprises (SME) and residential customers have been bandwidth-constrained, creating a 'bandwidth bottleneck'.

Typically, only larger enterprises can afford to pay the $1000+ per month that it costs to lease a 45 mbps connection. Purchasing T1s at $300 per month is an option for some medium-sized enterprises, but even today most small businesses and residential customers are left with few choices beyond dial-up Internet access. Where available, broadband internet access over DSL and cable modem offers a more affordable solution for data, but these technologies are difficult and time-consuming to provision, bandwidth is limited by distance and the quality of existing wiring, and voice services have yet to be widely implemented over these technologies. Most importantly, in rural and underserved markets, these wireline choices are simply not available as an option.

A wireless MAN based on the WiMAX–air interface standard is configured in much the same way as a traditional cellular network with strategically located base stations using a point-to-multipoint architecture to deliver services over a radius up to several kilometres depending on frequency, transmit power and receiver sensitivity. In areas with high population densities the range will generally be capacity-limited rather than range limited due to limitation in the amount of available spectrum. The base stations are typically backhauled to the core network by means of fibre or point-to-point microwave links to available fibre nodes or via leased lines from an incumbent wireline operator.

The range and NLOS capability makes the technology equally attractive and cost-effective in a wide variety of environments. The technology was envisioned from the beginning as a means to provide wireless last-mile broadband access in the MAN with performance and services comparable to or better than traditional DSL, Cable or T1/E1 leased-line services.

5.2   LAST-MILE HIGH-SPEED INTERNET ACCESS OR WIRELESS DSL

DSL operators, who initially focused their deployments in densely populated urban and metropolitan areas, are now faced with the challenge to provide broadband services in suburban and rural areas where new markets are quickly taking root. Governments are prioritizing broadband as a key political objective for all citizens to overcome the 'broadband gap' also known as the 'digital divide' (Figure 5.3).

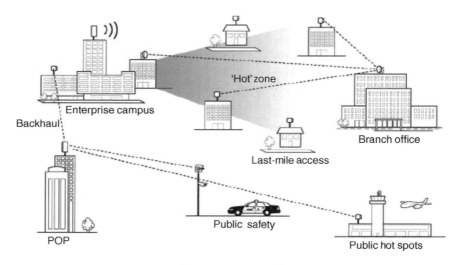

Figure 5.3 Last-mile

Large and Medium-sized Enterprises

Broadband Internet connectivity is mission critical for many businesses, to the extent that these organizations may actually relocate to areas where the service is available. In today's market, only 5 % of commercial structures worldwide are served by fibre networks, the main method for the largest enterprises to access broadband, multimedia data services. In the wired world, these networks are extended to the business or residence via cable or DSL, both expensive options because of the infrastructure changes required. DSL typically operates at 128 kbps to 1.5 mbps, slower on the upstream. Further, local exchange carriers have been known to take three months or more to provision a T1 line for a business customer, if the service is not already available in the building.

Older buildings in metropolitan areas can present a tangle of wires that makes it difficult to deploy broadband connections to selected business tenants. Enterprises can use WiMAX instead of T1 for about 10 % of the cost. IEEE 802.16a wireless technology enables a service provider to provision service with a speed comparable to a wired one. In addition, the range of IEEE 802.16 solutions, the absence of a line-of-sight requirement, high bandwidth, and the inherent flexibility and low cost help to overcome the limitations of traditional wired and proprietary wireless technologies.

Small and Medium-sized Businesses

This market segment is very often underserved in areas other than the highly competitive urban environments. For many small businesses that are out of reach of DSL or not part of the residential cable infrastructure, IEEE 802.16 represents an easy, affordable way to get connected to broadband. The WiMAX technology can cost-effectively meet the requirements of small- and medium-sized businesses in low-density environments and can also provide a cost-effective alternative in urban areas competing with DSL and leased-line services.

Residential and SoHo High-speed Internet Access

A low-cost alternative could end the war between the cable and ADSL operators and really make the broadband home revolution happen (Figure 5.4). Today this market segment is primarily dependent on the availability of DSL or cable. In some areas the available services may not

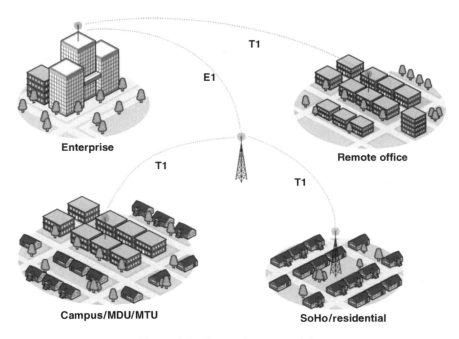

Figure 5.4   Enterprise connectivity

meet customer expectations for performance or reliability and/or are too expensive. In many rural areas residential customers are limited to low-speed dial-up services. In developing countries there are many regions with no available means of Internet access. The analysis will show that the WiMAX technology will enable an operator to economically address this market segment and have a winning business case under a variety of demographic conditions.

Underserved Areas

The most lucrative market for the proprietary BWA vendors has been remote regions, especially in developing countries but also in rural areas of the USA, where there is no wired or cellular infrastructure nor the will or cash to invest in building it. The main alternative to BWA in this market is satellite. Still early in its lifecycle – and potentially a powerful technology to integrate with WiMAX – satellite has severe limitations of upstream bandwidth, spectrum availability and also suffers from high latency.

Underserved markets include rural towns, and even newer suburban developments. An underserved area is any place where, for mostly economic reasons, high-speed wireline or wireless infrastructure was never constructed.

The main reason any market remains underserved is because it is difficult to develop a solid business plan for offering services there. Fortunately for all involved, public funding is playing a role in filling that gap.

Topography, bandwidth needs and finances influence whether a community gets broadband service and the best technology to use. Broadband wireless is the pre-eminent solution right now because it is the easiest to deploy and prices have come down.

Broadband on Demand

One aspect of the existing IEEE 802.16a standard that will make it attractive to service providers and end customers alike is its provision for multiple service levels. Thus, for example, the shared data rate of up to 75 mbps that is provided by a single base station can support the 'committed information rate' to business customers of a guaranteed

2 mbps (equivalent to a E1), as well as 'best-effort' non-guaranteed 128 kbps service to residential customers.

Depending upon regional demand, it should be possible for providers to offer a wide variety of standard and custom service offerings. By providing flexible service and rate structures to its customers, a WiMAX provider can appeal to a wide variety of needs by means of a single distribution point.

The key parameters of WiMAX receiving attention are concerned with its capability to provide differential services. Quality of service enables NLOS operation without severe distortion of the signal from buildings, weather and vehicles. It also supports intelligent prioritization of different forms of traffic according to its urgency.

MAC provide for differentiated QoS to support the different needs of different applications. For instance, voice and video require low latency but tolerate some error rate, while most data applications must be error-free, but can cope with latency. The standard accommodates these different transmissions by using appropriate features in the MAC layer, which is more efficient than doing so in layers of control overlaid on the MAC.

Many systems in the past decade have involved fixed modulation, offering a tradeoff between higher-order modulation for high data rates, but requiring optimal links, or more robust lower orders that will only operate at low data rates.

IEEE 802.16 supports adaptive modulation, balancing different data rates and link quality and adjusting the modulation method almost instantaneously for optimum data transfer and to make most efficient use of bandwidth. For rural areas, where the distances between customers are large, 'adaptive modulation' allows it to automatically increase effective range where necessary, at the cost of decreasing throughput. Higher-order modulation (e.g. 64 QAM) provides high throughput at sub-maximum range, while lower-order modulation (e.g. 16 QAM) provides lower throughput at higher range, from the same base station.

The modulation scheme is dynamically assigned by the base station, depending on the distance to the client, as well as weather, signal interference and other transitory factors. This flexibility further enables service providers to tailor the reach of the technology to the needs of individual distribution areas, allowing WiMAX service to be profitable in a wide variety of geographic and demographic areas.

5.3 BACKHAUL

Cellular Backhaul

Internet backbone providers in the USA are required to lease lines to third-party service providers, an arrangement that has tended to make wired backhaul relatively affordable. The result is that only about 20 % of cellular towers are backhauled wirelessly in the USA. With the potential removal of the leasing requirement by the FCC, US cellular service providers will also look to wireless backhaul as a more cost-effective alternative. The robust bandwidth of IEEE 802.16 makes it an excellent choice for backhaul for commercial enterprises such as hotspots as well as point-to-point backhaul applications (Figure 5.5). Also, with the WiMAX technology cellular operators will have the opportunity to lessen their independence on backhaul facilities leased from their competitors.

In Europe, where it is less common for local exchange carriers to lease their lines to competitive third parties, service providers need affordable alternatives. Subsequently, wireless backhaul is used in approximately 80 % of European cellular towers. Here the use of point-to-point microwave is more prevalent for mobile backhaul, but WiMAX can

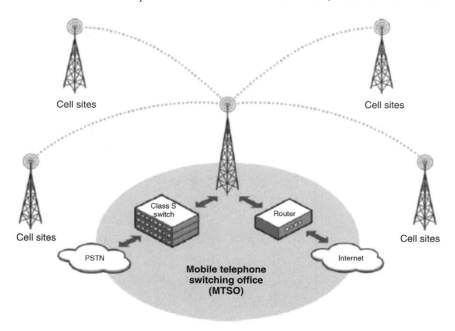

Figure 5.5   Cellular backhaul

still play a role in enabling mobile operators to cost-effectively increase backhaul capacity using WiMAX as an overlay network. This overlay approach will enable mobile operators to add the capacity required to support the wide range of new mobile services they plan to offer without the risk of disrupting existing services. In many cases this application will be best addressed through the use of IEEE 802.16-based point-to-point links sharing the PMP infrastructure. Some salient points about WiMAX use as cellular backhaul are:

- high-capacity backhaul;
- multiple cell sites are served;
- there is capacity to expand for future mobile services;
- it is a lower cost solution than traditional landline backhaul.

Clustered Wi-Fi Hotspot

Wi-Fi hotspots are being installed worldwide at a rapid pace. Wi-Fi hotspot operators may be able to build a spot for a few thousand dollars' worth of equipment, but then they need to anchor it to the public network, and this is normally done with expensive T1 or DSL (Figure 5.6).

The IEEE 802.11 standards were designed for unwiring the local area network (LAN); hence, their use in metro-access applications is facing many issues and challenges. Some of these challenges are non-standard

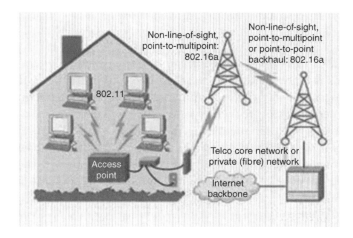

Figure 5.6 Wi-Fi backhaul

wireless inter-AP communication, the lack of capability to offer economic QoS hence voice and multimedia applications, and high cost of backhaul due to the use of wires, optics or other proprietary technologies.

Wi-Fi provides the certification for IEEE 802.11 client-to-AP communications. However, implementations of AP-to-AP and AP-to-service providers (that is, backhaul applications), which are typically needed for wireless last-mile and hot-zone coverage, are still proprietary, thus providing little or no interoperability.

The biggest obstacle for continued hotspot growth, however, is the availability of high-capacity, cost-effective backhaul solutions. This application can also be addressed with WiMAX technology. WiMAX backhaul could significantly reduce hotspot costs and, with nomadic capability, WiMAX could also fill in the coverage gaps between Wi-Fi hotspot coverage areas.

Last-mile broadband wireless access using WiMAX can help to accelerate the deployment of IEEE 802.11 hotspots and home/small office wireless LANs, especially in those areas not served by cable or DSL or in areas where the local telephone company may have a long lead time for provisioning broadband service.

5.4   THE RURAL BROADBAND PROBLEM

The conventional view today is that rural broadband is a problem. This arises from three commonly held propositions.

- *Rural broadband is necessarily more expensive than in urban areas.* If this is true, then these costs are likely to put off consumers and business, causing them to purchase only in small numbers. Fear of low demand may reduce investor and service provider confidence, deterring them from entering the market.
- *The market will not of itself meet the need.* If the market fails on its own to serve the need for rural broadband, then this will perpetuate the mounting 'digital divide' between rural and urban communities.
- *Some form of subsidy or other intervention is required.* If intervention is indeed necessary, then it may take various possible forms. A measure of non-commercial, that is subsidized, provision

is one approach. Innovative use of public–private partnerships is another.

A potential strategy for public sector broadband users is 'demand aggregation'. Public sector actors promote broadband services through their own concerted and coordinated demand, instead of fragmented, go-it-alone approaches. All schemes obviously depend on well-informed policy for their economy, efficiency and fairness.

The Costs of Rural Broadband

Rural broadband is generally believed to be more expensive than urban broadband for three reasons: distance, remoteness and scale economies. Nonetheless, it is worth bearing in mind the power of technological development to contest all these.

Rural dwellings and businesses are normally further from the point of supply of a utility service than their urban counterparts. The point of supply for rural broadband, or 'point of presence', is typically a local exchange building or radio base station. Many solutions, especially the cheapest, operate only up to modest distances. Limited ranges preclude application for many rural customers.

Broadband services depend not only on the last-mile supply, the access to the customer, but also on interconnection from the local point of presence to a high-capacity backbone optical network. While backbone networks provide plentiful high bandwidths very cheaply, they are only cheap when their capacity is filled. Such networks, therefore, naturally serve continents, countries and cities, but rarely visit the rural areas. Remote communities must, therefore, bear extra costs for distant connection between the local point of presence and a backbone network. The cost of this linkage, known as backhaul to a main network node, increases with remoteness, but is small or minimal in the urban environment.

Finally, broadband technologies frequently depend on platforms having high basic costs but a capability to serve many, perhaps a few hundred or more, connections. There is thus often a scale economy that cannot be realized in a rural community, raising unit costs. Technology can play a major role here, since it may succeed over time in reducing the minimum operational size of a platform. This shifts the scale economy, making the technology available to a wider customer base.

The Solution

Wireless DSL (WDSL) offers an effective, complementary solution to wireline DSL, allowing DSL operators to provide broadband service to additional areas and populations that would otherwise remain outside the broadband loop. Wireless Internet technology based on IEEE 802.16 is also a natural choice for underserved rural and outlying areas with low population density. Government regulatory bodies are realizing the inherent worth in wireless technologies as a means for solving digital-divide challenges in the last mile and have accordingly initiated a deregulation process in recent years for both licensed and unlicensed bands to support this application. Recent technological advancements and the formation of a global standard and interoperability forum, WiMAX, set the stage for WDSL to take a significant role in the broadband market.

5.5   3 VS: VoIP, VPLS AND VIDEO

Wireless VoIP

While VoIP has been around for years, it has not been a viable alternative for most applications due to technology constraints. Recent technology advancements have dramatically improved quality and now VoIP service providers are positioned to offer an affordable alternative to traditional circuit-switched voice services for both businesses and consumers.

VoIP services differ from traditional voice services because the voice conversation is transmitted over a proprietary broadband network or the public Internet. This allows VoIP providers to bypass the expensive public-switched telephone network (PSTN) and use a single broadband connection to transmit both voice and data. This not only reduces costs for voice providers that can be passed onto customers, but also enables corporate telecom providers to layer features such as unified messaging and Web-based call control through the convergence of voice and data.

Wireless VoIP is a simple and cost-effective service which allows a subscriber to use VoIP services while on the move. This is possible because of WiMAX which can provide carrier-grade connectivity while being wireless. It brings together the economy and benefits of VoIP and flexibility of wireless technology.

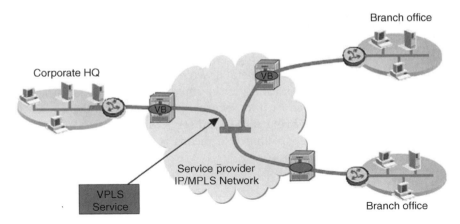

Figure 5.7   VPLS – virtual private LAN services

VPLS – Virtual Private LAN Services

Virtual Private LAN Services (VPLS) are a class of VPN that allows the connection of multiple sites in a single bridged domain over a provider-managed network. From the customer's perspective, it looks as if all sites are connected to a private LAN. VPLS provide another newer and rapidly growing enterprise data service that is replacing private lines and frame relay services (Figure 5.7). Over the next few years, VPLS will become the most popular WAN technology. WiMAX, owing to its QoS and security, provides an excellent network solution for VPN.

Video-on-Demand (VoD)

Video-on-demand, one of the most hyped technologies which never took off, may now get its due. With WiMAX has been found a technology which can make the base wider and price points better suited to the demands of customers. WiMAX can reach the masses at low cost, and hence more people who need services (unlike today when it is available to a few in city centres who have more economical ways to obtain video). Another interesting feature is that alternative videos and contents related to learning, training, etc. can become a revenue-generating mechanism due to obvious financial value. Other applications are as follows.

Automatic Teller Machines

The ability to provide ubiquitous coverage in a metropolitan area provides a tool for banks to instal low-cost ATMs across rural and suburban areas, which is a totally discounted possibility today because of the cost of satellite links and security issues with other modes of backhaul. WiMAX may bring ATMs and services kiosks to bank clients in suburban or rural areas. What that means is comfort for clients and enhanced business for banks.

Vehicular Data and Voice

WiMAX may be an innovation for fleet owners, logistic providers or logistic brokers, as they can find the location of their vehicles, their carriage capacity and amount of loading on real-time basis, which means better coordination for optimized carriage, unlike today when most carriers have to deal with low carriage on return trips. This may also help drivers and highway patrols to act quickly when facing adverse situations such as accidents or road blocks.

Online Gaming

If anything looked as appealing as pornography a few years ago, it was online gaming. With the emergence of this sector both in fixed and mobile forms globally, people without broadband are just waiting for technology to make access possible before more pervasive and faster growth. WiMAX will be the technology to provide the joysticks to rural and urban people at home or on the move.

Security and Surveillance Applications

Institutions of all kinds – from shopping centres to transportation systems to military bases – are being challenged to install video surveillance in areas that are too remote, too costly or physically impossible to reach with traditional cabling.

  WiMAX simply leaps over these barriers, allowing a virtually unlimited number of video surveillance cameras to be deployed quickly, easily

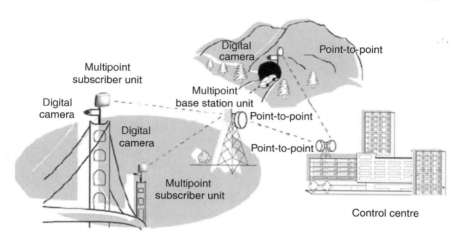

Figure 5.8 Security and surveillance applications

and cost-effectively in a new or expanded security system. High-resolution, real-time video from each security camera is transmitted directly to a WiMAX base station in the on-site security office or regional security centre. From here, the wireless network can remotely control the cameras (Figure 5.8).

Wireless video surveillance is a cost-effective, flexible and reliable tool for monitoring traffic, key roads, bridges, dams, offshore oil and gas, military installations, perimeter, borders and many more critical locations. Wireless video surveillance can also be used for special events as backhaul is easy and not time-consuming.

Further support for nomadic services and the ability to provide ubiquitous coverage in a metropolitan area provides a tool for law enforcement, fire protection and other public safety organizations, enabling them to maintain critical communications under a variety of adverse conditions. Private networks for industrial complexes, universities and other campus type environments also represent a potential business opportunity for WiMAX.

Because of its flexibility, WiMAX can provide a wide range of options from economical solutions for campus and mall security to mission-critical regional homeland security systems spanning thousands of square miles.

Multimedia Communication

IP-based wireless broadband technology can play an important role in delivering multimedia communication, information and entertainment that subscribers are demanding, with convenient access at any time and any place. Video chat and video conferencing are two such services, but with different quality and features.

Sensor Networks

Most mesh network applications, especially in the commercial sector, focus on traditional PC-based computing. However, researchers are also interested in using mesh network technologies to create networks of autonomous sensors – small devices that can be installed in a variety of locations to provide readings on temperature, air quality and other factors.

By incorporating a wireless chipset with mesh networking software, these sensors can become network-aware. After they are installed and powered on, the sensors can join a mesh network and make their data accessible to others on the network. In many situations, both in buildings and outdoors, installing small mesh-enabled sensors in many locations will be far preferable to setting up network cabling to connect the sensors or (worse) manually collecting data from the sensors.

Telematics and Telemetry

Telematics, the combination of telecommunications and computing, is predicted to be the next growth area in automotive electronics.

The use of automotive telematics in 'e-vehicles' that have audio email and web-browsing, DVD, digital TV and radio, as well as route guidance and traffic avoidance information, is predicted to grow to more than 11 million subscribers by 2004 in the USA.

A related technology, telemetry does not have such predictions. The Formula 1 (F1) sport utilizes telemetry to beam data related to the engine and chassis to computers in the pit garage so that engineers can monitor that car's behaviour. Bidirectional telemetry, from car-to-pit and pit-to-car, was allowed for a short while a few years ago. Bidirectional telemetry enables teams to alter settings on the governing electronic control unit by radio signal, and this can mean the difference between victory and defeat. However, car-to-pit telemetry is currently banned.

Many More

Some more disruptive applications of WiMAX can be:

- Remote monitoring of patients' vital signs in health-care facilities to provide continuous information and immediate response in the event of a patient crisis.
- Mobile transmission of maps, floor layouts and architectural drawings to assist fire-fighters and other response personnel in the rescue of individuals involved in emergency situations.
- Real-time monitoring, alerting and control in situations involving handling of hazardous materials.
- Wireless transmission of fingerprints, photographs, warrants and other images to and from law-enforcement field personnel.

6

WiMAX Impact

Wireless broadband access to the Internet has recently witnessed explosive growth. Much of this growth has come from the rise of wireless networks. Wireless networks today are being widely used in markets such as education, healthcare, manufacturing, retail, hospitality, government and transportation.

A new wireless network transition is gathering momentum in the shadow of the accelerating trend in wired broadband. The rapid growth of wireless infrastructure has generated an interesting set of problems for operators, users and regulators. Traditional operators have been forced to consider multiple business models to generate viable revenue streams.

Some of these models are similar to traditional ISP-type models that marked the spread of dial-up Internet access. Committed individuals, community-based networks and public networks, on the other hand, are taking a radically different approach by seeking to provide free wireless access.

6.1 BROADBAND FOR THE MASSES

Irrespective of whatever model is taken for bringing benefits to masses the basic service vision must have three key characteristics:

- High throughput and reach;
- Ubiquitous and predictable;
- Affordable and reliable.

The Business of WiMAX Deepak Pareek
© 2006 John Wiley & Sons, Ltd

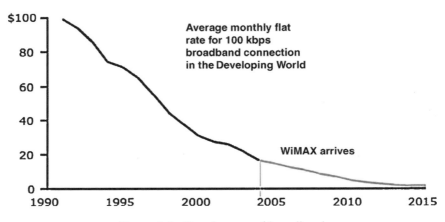

Average monthly flat
rate for 100 kbps
broadband connection
in the Developing World

WiMAX arrives

Figure 6.1   Trend – cost of broadband

WiMAX has a throughput of 75 Mbps and a range of 31 miles; future versions will be ubiquitous and they will be affordable. Having had a few false starts in an earlier incarnation as a purely fixed service, and having been catalysed by fundamental advances in wireless technology, the concept of true broadband, wide-area wireless is once again coming to the fore. Many industry players believe a new class of networks and services will be rolled out very broadly over the next few years. Leading indicators from recent commercial network deployments suggest that the prospects are indeed exciting: WiMAX will bring broadband to the masses (Figure 6.1).

6.2   AFFORDABLE BROADBAND

Once WiMAX Certified equipment is available from a number of suppliers, increased competition can occur, and with volumes of units shipped, more attractive price points can be reached. If WiMAX continues to gain support from industry, it can also provide broadband access in remote regions and developing parts of the world where basic voice or broadband access using fixed line service is not economically feasible. As WiMAX will be available as a system-on-chip, it will provide extraordinary benefits of cost and future innovation based on Moore's law.

6.3 MOORE MEETS MARCONI: WIRELESS APPLICATIONS

Moore's law started as a simple observation. It has since become a beacon for the electronics industry, guiding the efforts of chip developers and showing the rate of progress that must be maintained in order to remain competitive.

Now Moore's law is expanding to accommodate not just increased transistor count but also the rising complexity of silicon-based devices and the convergence of additional devices and technologies integrated onto the chip. With convergence imparting silicon power to communication, this is bringing about a new computing and communications landscape, making these technologies more affordable and widespread, and opening the door to broad new areas of innovation. This will ensure that Moore's law remains in effect for decades to come, through a combination of transistor count, complexity and convergence.

Researchers are creating on-chip smart radio circuits with built-in, reconfigurable wireless network hookups that offer always-on connections, plus the ability to switch automatically and transparently between wired and wireless networks. It is just one illustration of a basic principle: the principles of Moore's law to benefit entirely new arenas and enable expanded capabilities and performance (Figure 6.2).

Radio on Silicon

Before the advent of digital processing, radios were designed entirely from analogue circuitry. As advances in the cost and scale of CMOS technology provided digital processing power, digital signal processing

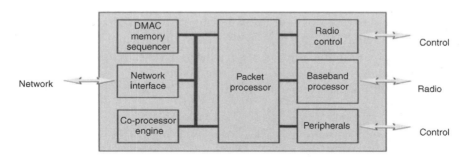

Figure 6.2   Radio on silicon

(DSP) began to play a major role in overall communication system design. Ever-improving DSP techniques have enabled improvements in communications consistent with the predictions of Moore's law.

Today, we are experiencing the power of DSP techniques through many wireless RF communication applications. Wireless wide area network (WWAN or cell phones), WLAN and wireless personal area networks (WPAN) all employ sophisticated communication techniques. Some of these techniques include complex modulation schemes, powerful new error correcting codes and decoding algorithms to combat the effects of channel fading.

All these techniques are being enabled, cost-effectively, by the increasing capabilities of digital processing. At the same time, CMOS technology and the effects of Moore's law have enabled digital devices to be produced in high volumes, again in a cost-effective manner, enabling larger markets.

Until recently, high-frequency wireless communications applications have used technology processes such as gallium-arsenide (GaAs) to obtain the performance needed from the RF analogue front end (AFE) circuits. Although these processes provide the functional performance required by radios today, they do not support the same cost/scalability economics of standard CMOS that is reflected by Moore's law.

The higher switching speeds that result from the smaller geometries being developed in CMOS are enabling the design of analogue circuits at very high frequencies, with very good gain and linearity. This new capability will allow analogue circuit designs to scale with the digital capabilities predicted by Moore's law. Analogue solutions implemented in CMOS will achieve high performance, functionality and bandwidth while maintaining low cost, small size, high quality and robust architecture across the wireless market.

The PC industry has always enjoyed the 18 month half-life trend, a corollary of Moore's law, unlike in communications. For example, PC prices have decreased by half every 18 months, but our communication costs have remained relatively stable over this period. Typically the 'half-life' of communication prices has been 5 years. Thankfully, all of that is about to change due to convergence. The new convergence industry will take advantage of traditional PC architectures, two examples being

- CPUs (e.g. Motorola PowerPC and Intel's Pentium);
- Bus [e.g. PCI, Universal Serial Bus (USB) and IEEE 1394, Firewire].

Consequently, the same 'price erosion trends' will be observed, and Moore's law will apply to convergence networks as well. As routers evolve to multiservice and proprietary hardware migrates to PC-based hardware implementations, and as corporate networking solutions move to the commodity market, the benefits will be reaped by the end user. With the communications industry opening up to new providers, the market will become increasingly competitive. Consumers will soon enjoy this cost cutting, and the aggressively competitive market in our communication hardware and software (Figure 6.3).

System on Chip (SoC)

The future digital lifestyle, which can be best summarized as anytime, anywhere, any device and any content, has raised the need for the development of system with broadband connectivity and a smaller, portable device with low power consumption, wireless connectivity and low cost.

Such a demand requires highly integrated and multifunctional personal devices or consumer electronics (CE), so-called 'embedded systems' in consumer space. While SoC technology is rapidly rising to be the most essential technology for future CE, the nature of the CE industry demands that the SoC R&D paradigm be changed.

These devices are multifunctional converged equipment, consisting of various components including modem, video, 3D, CPU/DSP, bus and software. Ignoring other facets like digital camera or video, these devices are capable of using multiple technologies for broadband wireless access.

It is composed of a network layer, I/O (input/output) layer and a processor. What previously were separate components are now merging into one chip, thanks to the SoC technology. A processor platform consists of a CPU, DSP, accelerator, SoC bus and memory (which will not be discussed in this context).

With the growing demand for the processor platform to support multiple communication standards (such as CDMA, W-CDMA, GSM/GPRS, WLAN 802.11/16/20 and UWB) at a single terminal and provide standard OS (operating system) support, a combination of an efficient processor architecture, high-performance bus and low power consumption is necessary. Thus, the adoption of a programmable, configurable 'CPU + DSP + accelerator' architecture is increasingly becoming favourable.

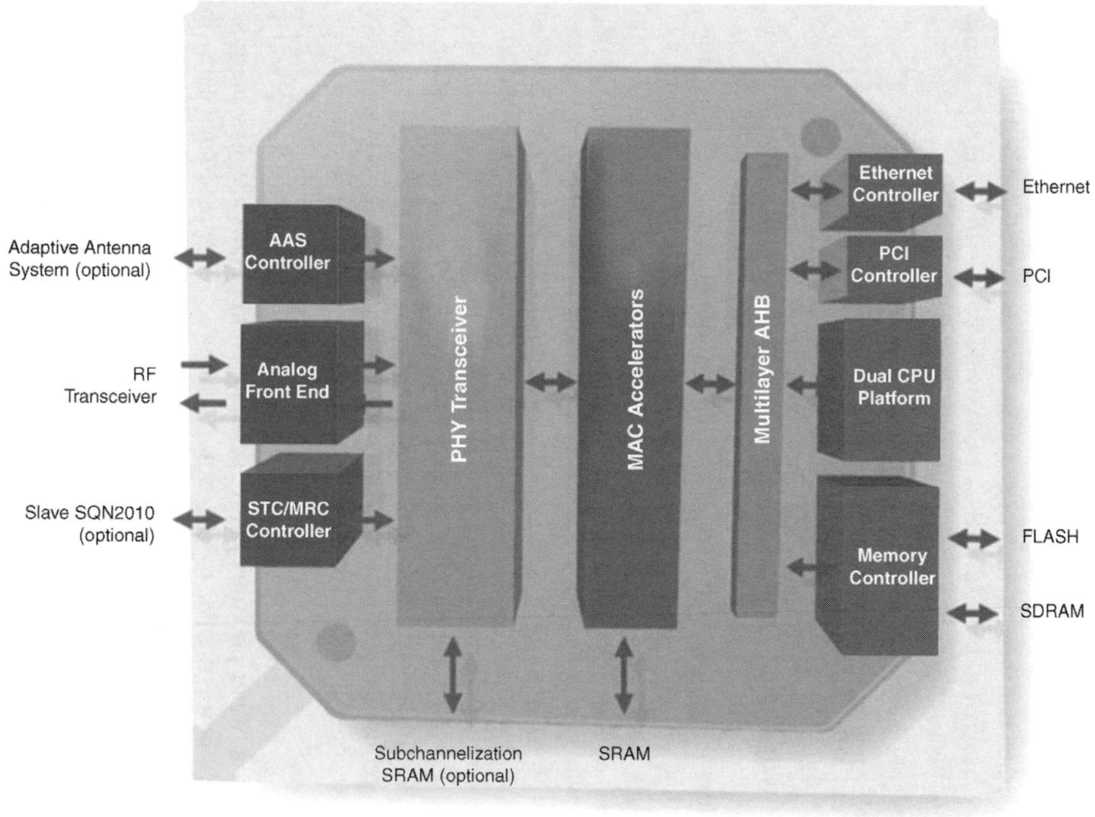

Figure 6.3   WiMAX SoC

6.4   EXPANDING COMPETITION: WiMAX

WiMAX is the first widely backed wireless standard that is both technically capable and has sufficient industry support to disrupt the telecommunications landscape. It is potent enough to turn on its head the connectivity stranglehold of incumbent telecommunication operators.

WiMAX provides an economically viable broadband wireless access technology, and provides extraordinary value to service providers as well as end users. It serves new entrants as well as dominant national incumbent operators with access and backbone infrastructure (Figure 6.4).

DSL and Cable

WiMAX deployment as a last-mile service not only serves residential and enterprise users but also as a backhaul for Wi-Fi hotspots and between the conventional cell towers. There are different opinions on whether BWA will be successful as a last-mile service. Our study lays the groundwork for deploying future BWA systems based on WiMAX Certified products. There are challenges in deploying WiMAX, but it has huge potential to compete on a cost-per-megabyte level with cable and DSL, if both engineering and economics are carefully applied.

We mainly focus on backhauling and tower leasing, by exploring several opportunities for significant cost savings, like aggregating the backhaul traffic and optimal use of tower space. WiMAX is all about

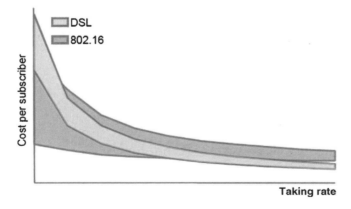

Figure 6.4   WiMAX vs DSL

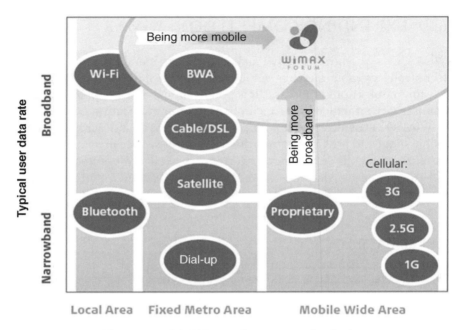

Figure 6.5 WiMAX vs other access technologies

delivering broadband wireless access to the masses. It represents an inexpensive alternative to DSL and cable broadband access. The installation costs for a wireless infrastructure based on 802.16 are far less than those for today's wired solutions, which often require cables to be laid and buildings and streets to be ripped up.

For this reason, WiMAX makes an attractive solution for providing the last-mile connection in wireless metropolitan area networks. Many countries in Europe have long-established copper networks where up to 40% of the fixed-line subscribers cannot benefit from DSL because of distance limitations or the use of pair-gain technology (using network amplifier systems to increase the strength of signals on paired wires; Figure 6.5).

Broadband Wireless Access Technologies

The history of broadband wireless has been largely one of disappointment. Pioneers like Teligent, Nextlink and Winstar entered the market in the late 1990s with networks based on cost-effective local multipoint distribution system (LMDS), but they played safe and stayed in over-served metropolitan areas of the USA rather than remote regions, and, having paid huge federal fees for their licences, all three companies filed for bankruptcy.

Carriers such as MCI and Sprint invested in an alternative, multichannel multipoint distribution system (MMDS) but failed to gain significant market momentum. Hence there was excitement around Wi-Fi hotspots, hotzones and community networks, but coverage scalability beyond a few nodes was hard to achieve without performance failures.

Since 2000, the BWA industry has matured. The big differences from the past are that:

- BWA has survived the recession;
- market consolidation has begun;
- operators are growing with excellent gross margins;
- improved equipment is available;
- costs have declined and financing is available;
- uptime has improved;
- WiMAX is coming;
- the public perception of wireless has improved;
- Wi-Fi hype has increased awareness of BWA/WISP opportunities;
- there are proven track record and referrals.

This has led to industry opportunities:

- re-emergence of interest in BWA funding;
- few regional players are poised for rapid growth with strong operations and management expertise;
- growth potential of WiMAX standard.

However, the BWA market is still developing:

- there is no successful national broadband wireless carrier, even today;
- major telco carriers are still at the residential trials stage;
- the market is highly fragmented, primarily with just local operators;
- small players are hitting the limits of management expertise and lack capital to expand.

BWA did not achieved widespread acceptance because of high customer acquisition costs:

- expensive CPE/SS – ~$500–1000;
- expensive LOS installation – requires truck roll/technician;
- unable to cost-effectively support QoS for more advanced services (VoIP);
- NLOS for indoor CPE.

Enter WiMAX, promising a lower cost backhaul for these hotspots than T1 and the option of a mesh network topology, as well as being a wireless extension to cable, fibre and DSL for last-mile connection. WiMAX is not a new technology. It is a more innovative and commercially viable adaptation of a technology already used to deliver broadband wireless services in proprietary installations around the globe. Wireless broadband access systems are already deployed in more than 125 countries.

What differentiates 802.16 from earlier BWA iterations is standardization. In these earlier solutions, the chipsets were custom-built for each broadband wireless access vendor, requiring a great deal of time and cost. Intel, Fujitsu and others would like to bring economies of scale to broadband wireless cost savings that would go a long way towards creating a larger market.

The Industry is waiting for a WiMAX opportunity:

- confidence;
- costs;
- bandwidth;
- WiMAX will change the model (Table 6.1).

WiMAX has emerged and everybody is getting excited again. It is no accident that the name sounds like 'Wi-Fi'. Although the two technologies are different, WiMAX proponents would like nothing better than to emulate and even overtake the success of 802.11. Several factors foreshadow a smoother acceptance of WiMAX than previous wireless technologies.

WiMAX has been designed so that it does not require LOS, and should allow higher speed downloads over much longer ranges than Wi-Fi. In part, this is because devices will support certain licensed spectrum bands, enabling them to transmit at higher power levels than would be possible using the Wi-Fi unlicensed spectrum. The description of WiMAX as 'Wi-Fi on steroids' is already on its way to becoming a cliché.

Table 6.1 Major wireless technologies for expansion of Internet access

Technology	Description	Strength	Weaknesses
VSAT	Small satellite terminal that can be used for one-way and/or two-way interactive communications via satellite	Can be installed virtually anywhere. High bandwidth availability	High initial costs. Difficult installation
Wireless local loop	Use of wireless radio signals to provide either voice or both voice and data services to fixed-point subscribers (primarily residential) who are not currently served by landlines. Broadband application examples include CoreDECT	Infrastructure built and maintained by telco normally for voice. Deployment quick and easy, reasonable data rates. Inexpensive, proven in rural settings. Good range	Costs do not scale well. Limited bandwidth technology is somewhat immature. Proprietary systems. Requires licences
MMDS/LMDS	MMDS is generally used to refer to fixed microwave data distribution systems below about 10 GHz. LMDS is generally used to refer to microwave data distribution systems operating above 10 GHz, generally in the range 24–32 GHz (10 mm wavelength) making them only useful for line-of-sight (LOS) communications.	Wide area coverage (20–40 km). Higher data rates	Should be carefully engineered over a line-of-sight and utilize external antennas carefully aimed and set up
Wi-Fi	Wi-Fi (802.11x) is a relatively mature set of IEEE standards for wireless networking of a LAN	Ideal for distribution within a small geographic area (like a village). Relatively low-cost. No communications infrastructure required. Can use special hardware to extend range. Mature technology	Limited range (200 m) for standard hardware. Crossing long distances requires special hardware at higher cost

Table 6.1 (*Continued*)

Technology	Description	Strength	Weaknesses
WiMAX	WiMAX (802.16x) defines family of wireless networking standards, defines a set of solutions for MAN with a range of up to 50 km	High data rates (up to 70 Mbps) Covers distances up to 30 km Easy to provision new service	Not yet widely available. Requires additional backhaul to feed wireless network. Uses licensed and unlicensed band, no mobility
Bluetooth	Bluetooth is an advanced protocol for the transmission of data at a rate of about 1.5 Mbps between portable devices in a secure and reliable manner.	Indoor or personal use does not require licensing	Limited distance. Lower data rate. Limited to commercial application
3G	3G stands for a third-generation mobile phone system which provides both voice and broadband data access. These include CDMA2000, W-CDMA, UMTS, GPRS and EDGE. GPRS and EDGE are actually 2.5G technologies because they are built upon existing 2G (second generation) GSM infrastructure.	Mobile extension. Good mobile data rates	The frequency bands (800, 900, 1800 and 2500 MHz) need licensing. Uptake has been limited. Expensive infrastructure
Mesh networks	Mesh networks are wireless networks composed of autonomous nodes that are able to self-organize. Each node is a wireless (often 802.11-based) radio unit that contains software that enables it to act as a mini-router. By creating a multihop network that spans many nodes, a mesh network can effectively extend the range of a traditional 802.11 network.	Extends the range of wireless technologies. Small scale-up increment. May be used to create a more robust distribution network	Relatively immature technology. Must be combined with other technologies

Surface Similarity   Some opportunities exist to leverage Wi-Fi chip technology for WiMAX devices; both technologies use orthogonal-frequency-division-multiplexing (OFDM) modulation. However, the two implementations are not identical. Because the two technologies operate in the same frequency neighbourhoods, RF specialists will have an opportunity to 'draft off' Wi-Fi work. However, the technologies require substantially different implementations. WiMAX radios require higher power because they must transmit over much longer distances than Wi-Fi radios. By extension, regulatory issues play more of a role. The fact that WiMAX operates in licensed as well as unlicensed swaths of spectrum represents another big contrast with Wi-Fi.

Another major point of contrast is the flexibility (or lack thereof) inherent in the standards. Whereas each 802.11 standard dictates one channel width and one frequency (either 2.4 or 5 GHz), 802.16a/REVd can work in a number of frequency bands and the available frequency can be sliced into a variety of channel widths. This flexibility explains widely varying numbers for the bandwidth and range capabilities of WiMAX systems.

In many ways, 802.16 is complementary to the local area IEEE standard, 802.11 or Wi-Fi. 802.16a provides a low cost way to backhaul Wi-Fi hotspots and WLAN points in businesses and homes, and as uptake of Wi-Fi increases, the requirement for this backhaul will grow too. Notably, this network can extend the Wi-Fi hotspot usage model to provide broader IP data service coverage and roaming that has so far eluded current 3G systems, due to system cost and complexity.

However, there is conflict too. WiMAX makes redundant the efforts of Wi-Fi specialists to extend the reach of their favourite technology and also places 802.11 into a far smaller role than its supporters have carved out for it, often unrealistically. This is the opportunity for wireless technologies finally to grow up and offer the speed, multimedia support and ubiquity that Wi-Fi can never deliver.

The newer standard holds all the real power. By providing a backbone for hotspots, based on standards rather than the various proprietary WLAN expansion technologies out there, it makes the idea of a ubiquitous wireless network to rival cellular far more realistic than it ever was with Wi-Fi alone, despite the claims of the enthusiasts. The equipment makers are eyeing it keenly – amid all the doubts about the sustainability of the hotspot boom, anything that offers them a new product line and helps to preserve the interest in Wi-Fi is to be welcomed. 802.16 is a highly complex standard which contains from day one many of the features that are being retrofitted, with various degrees of clumsiness and baggage, into Wi-Fi, which was originally conceived as being very simple and is now taking on the burden of responsibility beyond its technological reach.

The 802.16-2004 standard specifies networks for the current fixed access market segment. The 802.16e amendment and the soon to be approved 802.16f and 802.16g task groups will amend the base specification to enable not just fixed, but also portable and mobile operations in frequency bands below 6 GHz.

802.16 is optimized to deliver high, bursty data rates to subscriber stations, but the sophisticated MAC architecture can simultaneously support real-time multimedia and isochronous applications such as VoIP as well. This means that IEEE 802.16 is uniquely positioned to extend broadband wireless beyond the limits of today's Wi-Fi systems, both in distance and in the ability to support applications requiring advanced QoS such as VoIP, streaming video and online gaming.

WiMAX has various features that make it suitable for the longer term, although some like QoS may be incorporated into 802.11, which has failed to come up with specifications of its own in this area with any credibility. The 802.16a specification uses various PHY variants, but the dominant one is a 256-point orthogonal frequency division multiplexed (OFDM) carrier technology, giving it greater range than WLANs, which are based on 64-point OFDM. Another key difference of 802.16 is its use of time slots, allowing greater spectral efficiency for QoS capabilities.

Unlike the horror show that Wi-Fi went through with security, 802.16d WiMAX will use Triple Data Encryption Standard and Advanced Encryption Standard from its inception, a requirement for vendors doing business with many government organizations.

Systems based on the mobile version of the standard, which should ship towards the end of next year, will be able to achieve long distance wireless networking and will have far greater potential than Wi-Fi hotspots to provide ubiquitous coverage to rival that of the cellular network. Whether used directly or as backhaul for Wi-Fi, WiMAX fills the gaps in the hotspot system, and possibly enables it to challenge the cellular network as it cannot realistically do right now. In the end, the technologies will coexist in a creative way, with WiMAX increasingly the dominant partner, and the non-standard alternatives will fade into the background.

On one hand we have WiMAX, on the other, 802.20, nicknamed 'Mobile-Fi', the first standard to be specifically designed from the outset to carry native IP traffic for fully mobile broadband access. According to the latest Revision 13 requirements specification, 802.20 is a 'specification of physical and medium-access-control layers of an air interface for interoperable mobile broadband wireless access systems, operating in licensed bands below 3.5 GHz, optimized for IP data transport, with peak data rates per user in excess of 1 Mbps over distances of about 15 km. It supports various vehicular mobility classes up to 250 km/h in a MAN environment and targets spectral efficiencies, sustained user data rates, and numbers of active users that are all significantly higher than achieved by existing mobile systems.

This makes it lower powered than WiMAX but more intrinsically mobile, offering latency of 10 ms even in a fast moving vehicle, compared with 500 ms for 3G. Although the outcome of this battle is uncertain, delivery systems are sure to evolve that extend network access far beyond the limitations of Ethernet's original wired model.

However, 802.20 has three critical weaknesses – WiMAX is starting to take on some of its remit; WiMAX has stronger and more aggressive support from key vendors; and the mobile operators, while relatively friendly towards 802.16, are hostile to 802.20.

WiMAX has a huge head start on Mobile-Fi; even its 'e' version is at least a year ahead of its rival, and the industry support behind it is gathering pace rapidly. Also, it is a technology that can be accommodated relatively easily by the mobile operators (Figure 6.6).

3G Cellular Technologies

Unlike the popular view that WiMAX does not compete with 3G cellular, or the more audacious view by cellular operators that WiMAX will be a non-starter or will be too late to topple or even

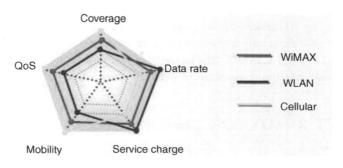

Figure 6.6 WiMAX, WLAN and 3G value analysis

impact 3G, WiMAX overlaps cellular technology significantly, and every cellular operator will have to consider WiMAX seriously. Further carrier WiMAX initiatives will not only impact service providers but are capable of disrupting the entire carrier ecosystem, including infrastructure and access device makers, phone vendors and even chip makers.

Every cellular operator (who does not want WiMAX to be a reason to go out of business) is going to have to consider WiMAX in their strategic planning. WiMAX is a serious threat to 3G because of its broadband capabilities, distance capabilities and its ability to support voice effectively with full QoS. This makes it an alternative to cellular in a way that Wi-Fi can never be, so that, while operators are integrating Wi-Fi into their offerings with some alacrity, looking to control both the licensed spectrum and the unlicensed hotspots, they will have more problems accommodating WiMAX.

WiMAX is likely to offer better performance than 3G where ubiquitous coverage and high mobility are not a priority. However, as with Wi-Fi, it will be better for them to cannibalize their own networks than let independents do it for them, especially as economics and performance demands force them to incorporate IP into their systems. Service providers that have 3G spectrum and services can offer an overlaid WiMAX broadband service targeted towards fixed, portable and nomadic subscribers with alternative devices such as laptops.

WiMAX could 'bridge the gap' between applications designed for high-capacity landline networks and the mobile broadband wireless networks. Handset makers such as Nokia will be banking on this as they develop smartphones that support WiMAX as well as 3G.

WiMAX can slash the single biggest cost of deployment for 3G: access charges for linking a cell to a local phone or cable network. A standards-based long distance technology will avoid many of the problems of high upfront costs, lack of roaming and unreliability that those ahead-of-their-time pioneers encountered, but it will still need to gain market share rapidly before 3G takes an unassailable hold. Given the current slow progress of 3G, especially in Europe, and the unusually streamlined process of commercializing WiMAX, the carriers are indulging in wishful thinking when they say nothing can catch up with cellular (Table 6.2).

6.5   GET READY FOR DISRUPTION

'802.16 is the most important thing since the Internet itself' – Intel.
The displacement of telegraphy by telephony offers insights into a process of innovation and product market competition that lies

Table 6.2 Comparison of 2.5 and 3G technologies and WiMAX

Metric	Cellular				WiMAX	
	Edge	HSPDA	$1 \times$ EVDO	802.16-2004	802.16e	
Technology family and modulation	TDMA GMSK and 8-PSK	WCDMA (5 MHz) QPSK and 16 QAM	CDMA2K QPSK and 16 QAM	OFDM/OFDMA QPSK, 16 QAM and 64 QAM	Scalable OFDMA QPSK, 16 QAM and 64 QAM	
Peak data rate	473 kbps	10.8 Mbps	2.4 Mbps	75 Mbps (20 MHz channel) 18 Mbps (5 MHz channel)	75 Mbps (max)	
Average user throughput	T-put <130 kbps	<750 kbps initially	<140 kbps	1–3 Mbps	80 % performance of fixed usage model	
Range outdoor (average cell)	2–10 km	2–10 km	2–10 km	2–10 km	2–7 km	
Channel BW	200 KHz	5 MHz	1.25 MHz	Scalable 1.5–20 MHz	Scalable 1.5–20 MHz	

at the heart of the next wave of growth in wireless. To grow
faster than the market a company needs to win new, and typically
more demanding, customers. This drives companies to work
hard to 'catch up' with the needs of these more demanding market
tiers.

Sustaining innovations are what enable organizations to appeal to
more demanding customer segments, and hence to grow. Sustaining
innovations can be incremental, year-by-year improvements, or they can
be leapfrog-the-competition breakthroughs. Either way, as companies
move along their sustaining trajectories, they inevitably 'overshoot' at
least some and sometimes much of the market, and as a result leave
behind those segments whose needs can be satisfied with lower performing and even inferior products.

It is this overshoot that makes disruption possible. Disruptive
innovations introduce products that are inferior to currently available
products – at least in terms of traditional performance measures.
However, they do offer other benefits. In general, disruptive innovations are simpler and more convenient to use, and less expensive than
the products and services that tend to dominate mainstream markets
(Figure 6.7).

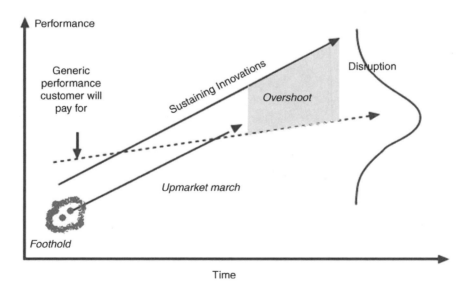

Figure 6.7 The theory of disruption

What is a 'Disruptive Technology'?

A disruptive technology is one which dramatically changes the way things are done. A disruptive technology has the following characteristics.

1. It is simple to use. WiMAX is optimized for NLOS and self-installing systems, which means that users do not need extra skills to make this system work. *System on Chip*

2. It is cheap to acquire and deploy. WiMAX is aided by SoCs, RFICs, interoperability and volume, which will drive system costs down to levels not seen in the past. Further, it does not need truck-rolls for installation.

   radio freq Integrated CCT

3. It is capable of targetting the masses. WiMAX provides optimum price-performance for BWA large wireless MAN deployments with hundreds of customers per base station and large ranges.

4. It is competitive against established products. WiMAX, more spectrally efficient, offers carrier-grade QOS, prioritization of voice/video and data, and higher network capacity.

5. It is commercialized in emerging markets. WiMAX is viable for deployment in markets where wired broadband is not cost-effective, i.e. underserved areas and the Developing World. It allows the rapid spread of broadband and higher speeds further away.

6. It is fast in technological progress. WiMAX forms the basis for future evolution to mobile BWA. It is the precursor for the next-generation mobile WiMAX standard (802.16e), fixed WiMAX available by the end of 2005, mobile available from mid-2006 in trials and commercial deployments in 2007 (Figure 6.8).

6.6 CATALYST TO ECONOMIC GROWTH

The Industrial Revolution and the subsequent mastering of the mechanical industry and of electricity production have granted prosperity and political clout to some countries. The Digital Revolution and subsequent mastering of the capability to control the technologies of information is a unique opportunity to earn economical and political rewards for the twenty-first century. This new century will see its full advent, building upon a universal access to two-way multimedia information (data,

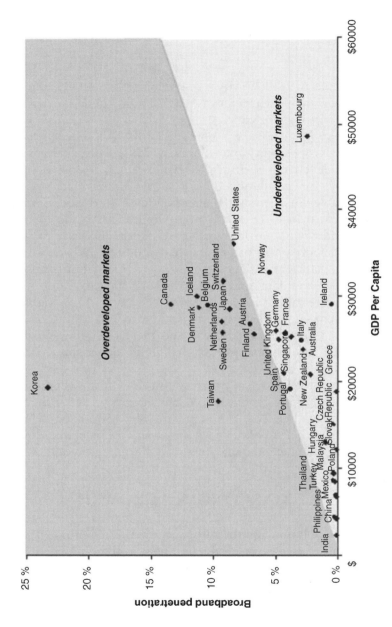

Figure 6.8 GDP impact on broadband penetration

sound, living images). Conversely, as in the preceding century, the laggards will fall behind in all domains, from culture to security, from independence to employment.

There are three main features in this revolution towards the digital age:

- *The revolution in everyday life – the application.* Instant communication of huge quantities of information (data, voice, image, sounds) between individuals and organizations will totally change the whole fabric of business and society.
- *The revolution in productivity tools – the technology.* Convergence of productivity tools with many different types of products/systems/services sharing common tools (the processor, the network, some software and middleware) has led to a radical shakeout in industry and services due to the fact that this communicating society is changing most of the current business models.
- *The revolution in availability and affordability of tools – access.* Affordability of these technologies is totally dependent upon a high level of advanced research and development driven by highly skilled personnel generating the choices for the future.

Telecommunications/ICT (information and communication technology), in the larger sense, is occupying an increasingly important role in society and economies. It is widely acknowledged to be a crucial driver of business growth and development, which in turn has the potential to influence other spheres – job creation, infrastructure, healthcare, education, etc. An adequate level of telecommunications and ICT infrastructure is an enabler for other businesses and an important precursor for economic development.

Information and Communication Technology

Telecommunications in the broad sense is a key enabler of a successful transition into the digital age, together with developments in other fields of human activity (culture, politics, education) and other industrial/technical domains (aeronautics and space, automotive, consumer goods etc.). Together they build on progress in basic technical disciplines (basic physics, microelectronics, software, hardware, networking) and integrate the results in terminal devices and infrastructures in order to produce services.

Like other domains, telecommunications face a dramatic increase in the pace of change in business and in technologies. Telecommunication plays a vital role in any nation's economy. Its application is ubiquitous in all sectors of an economy.

Telecommunication has evolved drastically since the invention of the telephone in the nineteenth century. It is now widely accepted that telecommunication is an important change agent. Research has confirmed a strong positive correlation between economic development and adequate telecommunication infrastructure. In developed as well as underdeveloped economies, telecommunication services are applicable to a wide range of economic production and distribution activities, delivery of social services and government administration.

Globally, the telecommunications sector has converged into an ICT sector. Telecommunication infrastructure serves as the medium for the flow of information in all forms, encompassing voice, data and video. In Africa, however, the level of telecommunication infrastructure is inadequate to service traditional telephony, let alone support information technology.

The importance of telecommunication cannot be overstated and is currently a major growth driver in many emerging economies that are trying to bridge both the technology gap and the development gap between them and the more advanced nations. As a result many countries have had to implement sweeping reforms in their telecommunication sector in the form of market liberalization and privatization of state-owned monopolies. Countries that have taken this action have witnessed substantial developments in their telecoms markets. Teledensity and quality of service have improved dramatically. Governments have earned huge revenues from telecommunication licensing exercises and these same authorities have freed themselves of the high cost of running inefficient state-owned telephone companies (Figure 6.9).

Broadband

Broadband networks are much faster than traditional dial-up connections. Broadband networks are fast enough to deliver a variety of simultaneous services, such as file transfer, streaming media (sound or video) and, most important, voice. Broadband networks can also deliver high-quality, uninterrupted service. Most networks are not made up of a single network technology and can include both wired and wireless broadband infrastructure components.

Teledensity

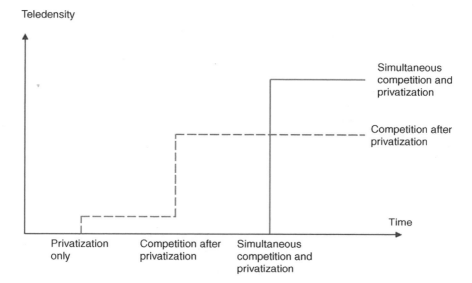

Figure 6.9 Impact of privatization and competition on teledensity

Serving the underserved with broadband wireless

Topography, bandwidth needs and finances influence whether a community gets broadband service and the best technology to use. For emerging and underdeveloped markets broadband wireless is the preeminent solution right now because it is the easiest one to deploy and the prices have come down. Further political and business landscape transformations are driving the global emergence of wireless broadband, including:

- regulatory changes;
- macroeconomic shifts;
- increased competition;
- technology innovations;
- cost and growth challenges.

With these changes comes greater competition for cheaper and more effective communications services that better address the business and economic needs of their markets. WiMAX is one such technology which provides maximum value, owing to it having the quality of wired and the cost and flexibility of wireless media.

WiMAX technology is the best hope for communities looking to add broadband access services, especially in regions abandoned by the big telcos. WiMAX has found a niche for itself in the 'emerging' market. With so many vendors looking to tap this niche, competition is bound to increase. Regardless of which vendor comes out on top, it is the millions of people in rural and developing markets who stand to gain the most from WiMAX.

Faced with solving real challenges, such as how to promote sustained economic development, ease road congestion in cities or improve in-patient care in hospitals, decision-makers in government, education and healthcare sectors are looking at WiMAX as a cost-effective tool to help promote social, economic and educational development. For example, small towns, struggling with technology deployment rates and the economics of traditional telecommunications networks, are investigating in delivering services over new WiMAX network environments. They are already driving investment in wireless infrastructures that deliver broadband access, and with WiMAX this is bound to become more attractive. Major cities worldwide are deploying applications and community services over wireless broadband networks to attract high-tech companies and foster economic revival. As more decision-makers recognize wireless broadband as a critical building block for development, the WiMAX proposition becomes more compelling (Figure 6.10).

6.7 THE ROAD AHEAD

Delivering Convergence

The conversion of telecommunication (telecom) networks and all forms of communication and information content to digital standards has created an electronic network infrastructure that facilitates the convergence of formerly discrete telecom services on a single telecom network. More recently, extended applications of IP have permitted the convergence of services on the Internet to include not only data, pictures, music and video, but also voice communication, including public voice services. VoIP is the latest major step in a convergence process that has been underway for three decades.

Convergence effects between telecommunication, information technology, media and entertainment (TIME) markets have led to additional value-added opportunities within these markets, allowing new intermediaries to enter the market. On the other hand, owing to these effects, traditional entry barriers to these markets have disappeared and new forms of competition and substitution for existing suppliers and products are emerging.

Figure 6.10 Convergence

This means that all types of services can be provided in an integrated manner over the Internet using IP. The Internet services in turn are provided over the digital network facilities of telecom operators. The convergence of telecom services using IP also completes a technical unbundling process that allows for a clear separation of facility network capacity from the services supplied over those facilities.

In the historic model of telephone service supply, specific services and facilities were integrated by technical design as both were supplied by telephone monopolies. IP has permitted a clear separation between network facilities and services, first for data, then pictures, audio, video and private voice communication, and now for public voice communication as well.

A more precise characterization of this more recent development in convergence using IP would be 'everything' over IP (EoIP). All forms of electronic communication now can be provided in an integrated fashion over a single network using IP. Although IP was developed for and initially applied on the Internet, the largest users of IP are telephone operators. They are in the process of converting their entire telecom systems to IP because of the enormous cost reductions to be achieved and the potential for providing new converged services in the future information economy, including e-commerce, e-government and other e-application services.

At the same time, the extended application of IP by Internet service providers (ISP) to include public voice services has opened a major new service opportunity for them, and introduced a significant new element of participation and competition in the supply of both public voice services and new converged services.

Operators that adapt their strategic business plan, considering the changing environment, with an early introduction of converged services will gain a competitive edge. Furthermore, the introduction of layered architecture will improve efficiency and flexibility and enable a smooth introduction of the IP multimedia subsystem (IMS), a cornerstone for efficient converged service offerings.

Definition of convergence

Traditionally, the term fixed-mobile convergence (FMC) has been used by the telecom industry when discussing the integration of wireline and wireless technologies. However, it is not just about this particular kind of convergence; it is also about convergence between media, datacom and telecommunication industries.

End-user expectations

User needs based on changing end-user behaviours and modern user expectations lead to three areas for further consideration, all of them closely related to convergence:

- *Convenience and ease of use* – users expect similar user interfaces for most services without having to consider which network is used. Services will be adapted to the device and access characteristics used, including simplified processes for identification and payments, as well as the ability to control cost.
- *Always best connected* – users expect to be able to connect anytime, anywhere – also when on the move – by their device of choice. Users also expect to be able to specify in each situation whether 'best' is defined by price or capability.
- *Reliability and security* – users expect reliability in all transactions, independent of access, and guaranteed connection quality. From a security point of view the user expects no viruses, worms or fraud, nobody listening in and the ability to know who requests a communication session.

Thus, one of the most profound changes in the way we look at convergence today is the increased end-user focus as a driver of convergence. Earlier, the focus was far more on operator and network efficiencies. These advantages, however, remain and have gained increased importance because of the new user and service behaviour together with maturing technology.

Network convergence

Until recently networks for wireless, wireline, data and cable TV services have existed in isolation. The next-generation solutions represent a more efficient way to build networks using a common multi-service layered architecture. The networks will have a layered structure with a service layer, a control layer, a backbone layer and access networks. Having one converged network for all access types is a significant benefit of layered architecture. This can improve service quality and allows the efficient introduction of new multimedia services based on IMS. Operators can increase network efficiency using optimized transport and coding solutions and will not need the over-capacity required when the networks are separated. Significant cost savings can arise from having one network with fewer nodes and lower operating costs. From an investment perspective, it is possible to optimize the use of control and media processing resources, hence reducing the need to replace technologies and the cost of network updates (Figure 6.11).

IP multimedia subsystem (IMS)

IMS provides a flexible architecture for the rapid deployment of innovative and sophisticated features. IMS focuses on introducing both a technical and commercial framework for a mobile operator to offer person-to-person services using a wide range of integrated media, voice, text, picture, video, etc. The standards have recommended the adoption of the session initiation protocol (SIP) as the service control protocol, and this will allow operators to offer multiple applications simultaneously over multiple access technologies such as GPRS or UMTS or ultimately other wireless or even fixed network technologies. The IMS standard will speed the adoption of IP-based multimedia on handsets, allowing users to communicate via voice, video or text via a single client on the handset. The vision for the IMS core network is

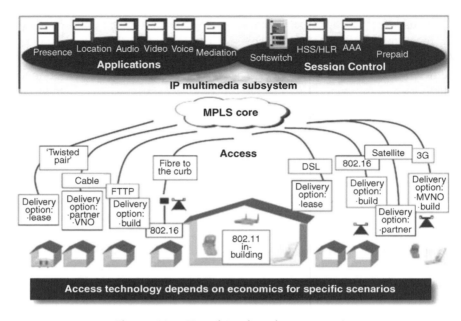

Figure 6.11   IP multimedia subsystem (IMS)

maximum flexibility and independence from the access technologies. This flexibility is accomplished, in part, via a separation of access, transport and control.

The control is further separated into media control, session control and application control. The radio access network provides the over-the-air connection from the user equipment to the core network. It also provides low level mobility management. The packet core network provides transport for the signalling and bearer, and high-level mobility management.

The IMS provides the control of applications, control of sessions and media conversion. Within the IMS, media control, session control and application control are separated in distinct entities. Some of the first applications expected to be launched using the standard will be push-to-talk over cellular (PoC), presence and instant messaging, and many other interactive applications eventually evolving to full fledged voice and VoIP. These applications can use a variety of basic network services offered by IMS like:

- session control services including subscription, registration, routing and roaming;
- combination of several different media bearer per session;

- central service-based charging;
- secure authentication and confidentiality based on the ISIM/USIM;
- quality of service support.

Besides these basic services, the IMS supports interworking with PSTN and CS domains for voice, and with corporate intranets, ISP networks and the Internet. Further, IMS is access-flexible and works together with any packet-based access network. This allows operators to leverage the IMS core infrastructure by using it not only for UMTS radio access, but also for GPRS, EDGE, TD-SCDMA, licence-free hotspot radio technologies (e.g. Wi-Fi) and wireline networks.

IMS is a key component of multiservice layered architecture. It is a subsystem supporting multimedia sessions, standardized by 3GPP and using SIP from IETF. IMS is a common foundation for fixed, mobile and enterprise services, delivering services over multiple accesses such as CDMA2000, WCDMA, GSM, fixed broadband and WLAN. Thus it is a cornerstone in a converged solution.

A converged network using IMS allows the following resources to be shared, regardless of service or access type.

- charging;
- presence;
- directory;
- group and list functions;
- provisioning;
- media handling;
- session control;
- operation and management.

In addition to making converged user services faster and easier to introduce, the common shared resources also increase operational efficiency in the network. The network evolution path is unique for each operator and depends on many factors including the business environment, cultural heritage, regulations, end-user behaviour and PC and mobile penetration rates. The strategic reaction of the incumbents of the TIME industries towards these developments is mostly to pursue cautious competition prevention strategies within these new fields of value creation according to the principle of 'cannibalize your systems before someone else does'. The transformation is usually done step-by-step towards the target network with an all-IP-solution based on IMS.

A cellular service provider can integrate WiMAX with an existing or planned mobile wireless solution in order to:

- take advantage of an existing subscriber management system;
- benefit from already installed network management capabilities;
- utilize a cellular deployment infrastructure already in place, such as existing cell towers;
- provide common and additional services to users' homes, including new multimedia services.

WiMAX: part of the converged network reality

WiMAX will be a key enabling technology for fixed/mobile network convergence. It can give service providers another cost-effective way to offer new high-value multimedia services to their subscribers. With the potential to deliver high data rates, along with mobility, it can support the sophisticated 'lifestyle' services that are increasingly in demand among consumers, along with the feature-rich voice and data services that enterprising customers require. Because it is an IP-based solution, it can be integrated with both wireline and 3G mobile networks.

This versatility opens up cost-effective new opportunities for extending bandwidth to customers in a wide range of locations and for delivering new revenue-generating services such as wireless video streaming. For wireline service providers, WiMAX will be another access technology that can be supported with the same operations and subscriber management infrastructure, while they take advantage of its capabilities to cover gaps in DSL service, extend their reach to underserved subscribers or provide last-100 meter access from fibre at the curb, minimizing the expense of fibre deployment.

For 3G wireless service providers, WiMAX offers a solution that is interoperable with their networks that can augment coverage and service to end-users. Service providers can take advantage of another of today's promising new technologies, leveraging it as a source for new revenues and delivering multimedia services wherever subscribers demand them (Figure 6.12).

Always Best Connected

Telecommunications is the merging of voice, data (WAN), LAN, video, image and wireless communications technologies with PC and micro-electronic technologies to facilitate communications between people or

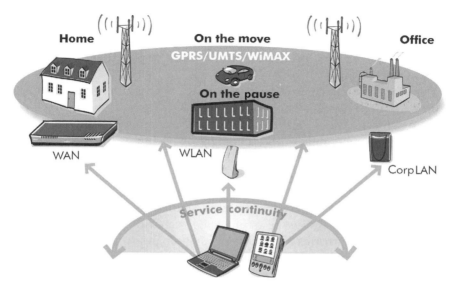

Figure 6.12 Always best connected

to deliver entertainment, information and other services to people. People around the globe are mobile, and they want all of their communications to support that mobility. Telecommunications represents a convergence of these technologies into networks and systems that serve people planetwide.

As wireless connectivity is becoming more widespread and more complex, the ability to provide service on the many levels available to wireless users using a variety of devices is also rapidly becoming much more complex. To accommodate these challenges and to face a future where there are no barriers to access using a handheld, portable or fixed device, engineers are investigating what measures are needed to create a 'universal always best-connected communicator,' a solution that is capable of communicating regardless of the connection options available to the user.

The WiMAX family of standards is still in its infancy, but the WiMAX Ecosystem is already addressing these challenges. WiMAX solutions can integrate with the standards-compliant IMS architecture, and it can utilize the home subscriber server (HSS) functionality in the IMS core to manage subscribers, simplifying authentication, authorization and roaming. The end result is a seamless, common experience as end-users move from a 3G mobile or other networks to a WiMAX network.

The Goal

The goal laid out for telecommunication industry is to keep end-users 'always best connected'.

Vendors and service providers envision always best connected coverage for multimode devices containing many wireless access technologies, e.g. Bluetooth, Wi-Fi, WiMAX, 2.5G and 3G mobile. This requires a level of interoperability and cooperation never seen before. Each technology has its own advantages and disadvantages, so being best connected means using the technology with the optimum signal and bandwidth characteristics for the application in use; and perhaps thereafter, the ability to handover between technologies.

To achieve the always best-connected scenario users will mix and match mobile platforms and wireless technologies to meet their unique requirements, enabling them to stay connected virtually anytime and anywhere.

Broadband wireless can reach the always best connected goal through the following scenario:

- all types of wireless networks will be deployed around the globe;
- Wi-Fi hotspots will proliferate in public places, businesses and homes;
- homes and businesses will add UWB (when available) for the fastest distribution of high-definition content;
- first-generation WiMAX technology will be broadly deployed to provide long-distance broadband connectivity for Wi-Fi hotspots, as well as cellular and enterprise backhaul;
- 802.16e WiMAX connectivity will be added in densely populated areas to provide a canopy of wireless broadband data access to mobile laptop users;
- innovations in 3G technologies will add ground-breaking data capabilities to mobile handset and handheld PC users.

Roaming and Seamless Handoff

Key to future impact of convergence is seamless handoff between disparate data networks, i.e. handoffs between Wi-Fi (802.11) networks and 3G mobile networks (CDMA2000® and UMTS), between 3G networks and 802.16 networks or between Wi-Fi (802.11) networks and 802.16 networks. The initial deployments of the 802.16d system will be a fixed point to fixed multipoint deployment. In this case, roaming will be handled from the fixed subscriber station distribution system and the users' communication platform, in much the same way as roaming from a DSL or cable modem is implemented. Once the 802.16e systems become available and certified, roaming will be accomplished in the same manner as roaming between 3G and Wi-Fi.

Challenges

Enabling such ubiquitously connected devices poses numerous difficult technology challenges. These include:

- Multiple radio integration and coordination – building the handset (or other device) begins with the challenge of integrating multiple radios.
- Intelligent networking, seamless roaming and handoff – users will expect to roam within and between networks, as they do with their cell phone.
- Power management – as handsets and other devices evolve to run more richer applications, power management will become an even greater challenge.
- Support for cross-network identity and authentication – providing a trusted, efficient and usage-model appropriate means of establishing identity is one of the key issues in cross-network connectivity.
- Support for rich media types – the addition of a high-bandwidth broadband wireless connection, such as WLAN or some of the forthcoming UMTS or EVDV/O cellular networks will open up new opportunities for the delivery of rich media to handheld devices.
- Flexible, powerful computing platform – the foundation of a universal communicator-class device must be a flexible, powerful, general-purpose processing platform.
- Overall device usability – the final challenge inherent in building a mixed-network device is usability (Figure 6.13).

Next-generation Network

New business initiatives and competitive pressures to become more efficient and productive are forcing enterprises to seek new types of

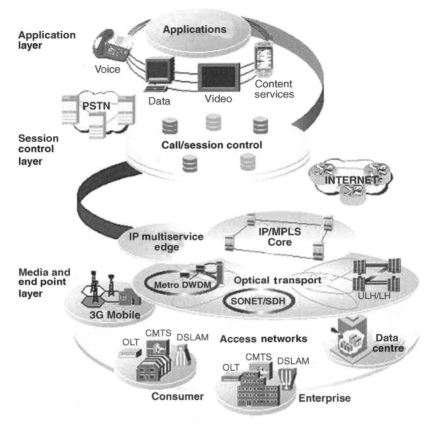

Figure 6.13 Next-generation network

network services. In particular, enterprises are showing significant interest in advanced data services, and are adopting such offerings as metro Ethernet services, IP VPNs, Layer 2 VPNs and VoIP technology as a replacement for traditional PBXs.

At the same time, carriers continue to rely heavily on the more traditional voice and data services that are well understood and widely deployed throughout their networks. Changes in customer demands are causing carriers to dramatically alter the way they design their networks. In addition, carrier requirements to cut their costs – both operational and capital expenses – are causing further reevaluation of networks.

The best network solutions for carriers are ones that meet their customers' needs for next-generation services and also address their own requirements for lower capital and operational costs. Enabling new services, while maintaining legacy services, is critical for their success in moving forward.

As voice and legacy data service revenue declines, carriers need new services to fill the revenue void. More importantly, enterprise customers are demanding new services from their service providers. If the incumbent provider does not have what a customer needs, they will spend their money with a new service provider that does offer it.

The requirement for service providers to support legacy services is clear from surveys of enterprise customer needs. Enterprises are not migrating to next-generation data services overnight. Legacy data services, such as frame relay and ATM, will continue for many years to come. Enterprise requirements for new services to interwork with legacy services underscore this reality.

In the environment just described, service providers face many risks in making their NGN network decisions, and in not making them. For instance, they face the risk of making a wrong technology decision. Technology investments are expensive, and carrier capital budgets are limited. If they place a strong bet on a data service that does not take off, they face the risk not only of wasting the capital and operational dollars invested in that service, but also of missing the 'next big thing' completely because they have drained their resources on the wrong choice.

Service providers also face the risk of making the right technology decision at the wrong time. It does service providers no good to be right about an enterprise technology migration but be two years too late to market, or two years too early. The key is to mitigate their network migration risks as much as possible. For the largest operators, a hybrid network architecture approach that preserves legacy services and legacy infrastructure investments while enabling the timely introduction and scaling of new services for enterprise customers presents the best network migration scenario.

Another NGN approach is to build an all-packet network. Driving the all-packet network opportunity is the fact that enterprises are migrating from circuit-to-packet services. VoIP is gaining momentum in the enterprise. TDM private lines, once the workhorse of the enterprise WAN, have been replaced in growth and significance by a host of data services, including emerging Ethernet and IP data services. Some service providers have built their networks around an all-packet approach, focusing primarily on various enterprise data services (with circuit emulation for TDM services). The challenge is that, as pointed out, enterprise networking needs are a mix of old services and newer services, voice and data, legacy data and next-generation data.

Key characteristics of NGNs are:

- Geographic transparency – boundaries are disappearing and economic benefits independent of service 'density' must be realized.
- Transport efficiencies – transport costs (price/b) are continuously declining, NGNs must share these efficiencies – for both bearer and signalling traffic
- Internet technology economics – leverage services and service delivery through the Internet, as well as the 'silicon economics' of Internet hardware (servers etc.) as memory and processor price/performance improve
- 'Old World' to 'New World' interoperability – existing PSTN infrastructure, and its associated investment must be fully utilized

Key considerations in the evolution

As carriers move towards solutions using new technologies and architectures, the success or failure of these solutions is dictated by many factors. In the case of the NGNs and 'deconstructed' switch/packet network replacing the circuit switch/TDM network, it is important that these key benefits be delivered:

- investment protection;
- operational and capital costs savings;
- carrier-grade reliability;
- improved service creation capabilities;
- scalability;
- improved product selection/choices.

Key drivers for NGNs

Today's circuit-switched architecture(s) have evolved over the last 100 years; however, this evolution compared with the rapid evolution seen in the PC and Internet worlds has occurred at a snail's pace. The lifecycle of equipment in the traditional telecommunications marketplace is measured in decades. Contrasting that with 'Moore's law' in the PC market place and 'Internet time' in the Internet communications space illustrates a key motivator for NGNs.

There are at least four key techno-economic drivers for NGNs:

- costs (capital and operational);
- price/performance;
- standards;
- speed of innovation and introduction of services.

With the dramatic hardware technology changes in the PC world (obsolescence typically in 18 months) and the rapid innovation within the Internet, both hardware and software products are on dramatically different price/performance curves from traditional telecommunications equipment. Industry analysts often quote IP network capital costs to be as little as 50 % of comparable switched network costs. Furthermore, the operational costs of combining (i.e. converging) the traditional disparate voice and data networks ostensibly could be in the 50 % range as well.

These cost advantages are key drivers for NGNs; however, they are not the only drivers. Historically, the traditional telecommunications networks depended on a selected group of vendors offering closed, proprietary solutions. These seldom interworked, except at clear lines of demarcation in standard interfaces. This market structure favoured vendors and allowed significant control of product evolution by these same vendors. Today's PC and Internet markets are based on 'consumer' market quantities (e.g. millions and millions) and rapidly evolving standards – but standards nonetheless. Since the NGNs leverage significant aspects of these two areas, the volume (that drive prices down) and the standards (that promote interoperability) form yet two more techno-economic reasons for NGN implementation.

The fourth and perhaps the most unproven reason for migration to NGN implementation is the ability of these new networks to support rapid introduction of new and different services. Consistent with the characteristics of Internet applications, however, this 'promise' has yet to be realized simply because NGNs are still in their infancy. If Internet technologies (NG HTML, DNS, LDAP etc.) are effectively used, rapid, innovative services may prove to be the most compelling reason for NGNs. Regardless of the level of NGN 'integration' into today's existing circuit switched networks, benefits will be realized. However, as the NGN technologies mature and are deployed, the more complete the move to NGNs, the more substantial will be the benefits.

PART Three

WiMAX Business

7

WiMAX Markets

Perhaps even before the phrase 'digital divide' was coined, the telecom industry was searching for a cost-effective way to bridge the gap between the 'haves' and 'have nots', making broadband services accessible to all.

Although broadband wireless is not an entirely new technology, the evolution to a standards-based, interoperable, carrier-class solution gives WiMAX the capability needed to drive wide-scale deployment.

WiMAX holds the promise of delivering broadband services on a ubiquitous scale to these underserved markets, which can be a rural landscape in advanced and developed countries like the USA or an urban setting in a least-developed nation like Ghana or Vanuatu.

7.1 MARKET DYNAMICS

WiMAX will succeed globally, albeit unevenly. It will succeed in every geographic market, but for different reasons. In emerging markets, operators are interested in using WiMAX for low-cost voice transport and delivery. In developed markets, WiMAX is all about broadband Internet access.

Overall, the markets without any fixed infrastructure pose the greatest opportunities. WiMAX will become a disruptively inexpensive means of delivering high-speed data. As the distinctions between fixed and mobile services blur, a chaotic mix of large, fixed and wireless providers will pursue WiMAX deployments.

The Business of WiMAX Deepak Pareek
© 2006 John Wiley & Sons, Ltd

The local and regional wireless ISPs are likely to be acquired as large carriers, particularly fixed carriers, turn their attention to rural areas and enterprise accounts. For example, in the USA we have seen ISPs compete successfully in moving enterprises from T1 lines to wireless lines; fixed carriers will eventually be forced to respond, either through similar deployments or through acquisition.

The reasons behind wireless deployments are as diverse as the wireless technologies being offered today. Each wireless technology is designed to serve a specific usage segment:

- personal area networks;
- local area networks;
- metropolitan area networks;
- wide area networks.

The requirements for each usage segment are based on a variety of variables, including:

- bandwidth needs;
- distance needs;
- power;
- user location;
- services offered;
- network ownership.

Optimized applications exist for each usage segment.

7.2 MARKET TYPES

Nearly two-thirds of the planet's people are in the underdeveloped world. Most companies are serving at best one-third of the world population and fiercely competing over saturated markets. Yet many corporate managers now realize that stepping up their company's presence in developing countries will be crucial to their long-term competitiveness and success.

For a large number of countries and for a large percentage of the world population, the telecommunication infrastructure is undeveloped; low-income economies in the world have an average of 1.97 main lines per 100 inhabitants and the lower-middle-income economies 9.17 main lines per 100 inhabitants in contrast to 47 main lines per 100 inhabitants in the developed world.

Economic development and growth requires a level of infrastructure which many countries in the world simply do not have. A certain level of economic advancement and industrial progress is generally required to alleviate poor social conditions for any meaningful period in the modern world. Where some basic infrastructure does exist and certain services do operate, the operation of such infrastructure is sometimes below the level necessary to enable sustainable socioeconomic development. Let us understand what is meant by developed and underdeveloped markets.

Developed Markets

Development in human society is a many-sided process. At the level of the individual, it implies increased skill and capacity, greater freedom, creativity, self-discipline, responsibility and material well-being. Some of these are virtually moral categories and are difficult to evaluate, depending as they do on the age in which one lives, one's class origins and one's personal code of what is right and what is wrong. However, what is indisputable is that the achievement of any of those aspects of personal development is very much tied in with the state of the society as a whole.

More often than not, the term 'development' is used in an exclusively economic sense – the justification being that the economy is itself an index of other social features. A society develops economically as its members increase jointly their capacity for dealing with the environment. Taking a long-term view, it can be said that there has been constant economic development within human society since the origins of man because man has increased enormously his capacity to win a living from nature.

Developed economies have certain characteristics. They are all industrialized. That is to say, the greater part of their working population is engaged in industry rather than agriculture, and most of their wealth comes out of mines, factories, etc. They have a high output of labour per person in industry because of their advanced technology and skills.

Underdeveloped Markets

Having discussed 'development', it makes it easier to comprehend the concept of underdevelopment. Obviously, underdevelopment is not absence of development because every population has developed in one way or another and to a greater or lesser extent. Underdevelopment makes sense only as a means of comparing levels of development. It is very much tied to the fact that human social development has been uneven and, from a strictly economic viewpoint, some human groups have advanced further by producing more and becoming wealthier.

After Britain had begun to move ahead of the rest of Europe in the eighteenth century, many Russians were very concerned about the fact that their country was 'backward' compared with England, France and Germany. Today our main preoccupation is with the differences in wealth between, on the one hand, Europe and North America and, on the other hand, Africa, Asia and Latin America. In comparison with the first, the second group can be said to be underdeveloped. At all times, therefore, one of the ideas behind underdevelopment is a comparative one.

The underdevelopment with which the world is now preoccupied is a product of capitalist, imperialist and colonialist exploitation. African and Asian societies were developing independently until they were taken over directly or indirectly by the capitalist powers of Europe. That is an integral part of underdevelopment in the contemporary sense; however, we will not pursue this discussion as the subject is out of scope of this book.

In some quarters, it has often been thought wise to substitute the term 'developing' for 'underdeveloped'. One of the reasons for so doing is to avoid any stigma which may be attached to the second term, which might be interpreted as meaning underdeveloped mentally, physically, morally or in any other respect. However, on the economic level, it is best to remain with the word 'underdeveloped' rather than 'developing' because the latter creates the impression that all the countries of Africa, Asia and Latin America are escaping from a state of economic backwardness relative to the industrial nations of the world.

Professional economists speak of the national income of countries and the national income per capita. These phrases have already become part of the layman's language by way of newspapers, and no explanation is needed here. The developed countries have per capita incomes several times higher than that of developing nations.

Characteristics of Underdeveloped

The main characteristics of developing countries are:

- Dualistic economies – they are composed of differential urban and rural markets, with the former far more advanced than the latter.

- General poverty and low income – the majority of the people in underdeveloped countries possess low levels of income because of their extremely low level of production.
- Underdeveloped natural resources – the national resources in an underdeveloped country are either unutilized or underutilized.
- Capital deficiency – generally, underdeveloped countries suffer from shortage of capital, which is responsible for the low per capita income in the economy.
- Excessive dependence on agriculture – most of the underdeveloped countries are predominantly agricultural and producing raw material products. In these countries, output and employment tend to be heavily concentrated in agriculture or mineral fields while manufacturing accounts for very little of the output and employment.
- Foreign trade orientation – generally, in these countries, imports are more than exports, leading to an unfavourable balance of payment situation.
- Existence of unemployment and disguised unemployment – in underdeveloped countries the phenomenon of disguised unemployment is very common. The phenomenon of disguised unemployment is confined to the agricultural sector. It is the result of excessive pressure of population on land, and most of the people lack alternative employment opportunities.
- Absence of enterprise and initiative – the absence of dynamic entrepreneurship and initiative is another common characteristic of underdeveloped countries.
- Technological backwardness – underdeveloped countries are backward in the level and character of economic performance as compared with their counterpart countries, which are the advanced countries of the world.
- Lack of infrastructure – in these countries the means of communication and transportation are not fully developed. Power generation is insufficient to meet the growing needs of the economy. Technical and professional facilities are poor.

These problems in underdeveloped countries constrain development.

The Ultimate Business Opportunity

Many companies are already targetting new customers and suppliers further down the economic pyramid. In doing so, they are gaining a first-mover advantage. Companies that learn to operate in new markets and thus improve their reputations will be at a competitive advantage as the countries become richer and more business opportunities emerge. Reputation benefits can also improve corporate relations with governments and communities and improve employee recruitment, retention and morale.

History shows that business, not government, develops a nation economically. Governments create the frameworks that encourage – or hinder – that development; but it is the private sector that generates entrepreneurship, creates employment and builds wealth. Companies, moving beyond conventional wisdom and working with new partners, have an unprecedented opportunity to help people to lift themselves out of poverty and into market economies.

These companies will be at the same time developing new, broad-based markets for their businesses. Business creates value by increasing revenues, lowering operational costs and improving productivity. It does so by growing new markets, tapping into new revenue streams and reducing costs through outsourcing and global supply chain management. It is increasingly looking at emerging economies and developing countries for such opportunities.

Business, working in a spirit of 'enlightened self-interest', can improve the developmental paths of billions of people, by facilitating their access to the marketplace, by finding new ways to address the needs of the poor and helping them into mainstream economic activity. A growing body of evidence indicates that intelligent engagement will also result in new revenue and profits.

Encouraging Trends and Drivers

Global trends are creating a favourable environment for companies to start engaging with the poor.

- *Many companies see a need to break out of mature market sectors –* forward-looking companies see the most attractive growth opportunities in emerging markets with young, dynamic populations and economies.
- *Framework conditions in many developing countries are improving* – countries across the globe are investing time and effort in strengthening their governance, legal structures and investment infrastructure. Progress is far from uniform, but there are signs of improvement in many countries.

- *Communications are faster and cheaper, making the world a smaller place* – lower communications and transportation costs allow more geographically dispersed production. This can allow companies to benefit from lower labour and material costs and encourage them to relocate part of their activities in developing countries.
- *Public expectations of corporations are changing* – communities and civil society increasingly expect companies to become involved with social issues. Many have already realized that it is better to do so proactively, in partnership with others, than reactively, as was sometimes the case with the need to mitigate corporations' environmental footprint.
- *New, and better, partners are available* – many multilateral organizations are experiencing their own far-reaching changes, driven by their need to become more self-sustaining and to improve their effectiveness. Today, many of these groups are prepared and are able to help companies operating in poor countries and poor neighbourhoods. They also understand how companies can help them realize their own goals of improved business opportunities in the developing world.
- *Aid and investment are beginning to reinforce one another* – the importance of foreign direct investment (FDI) as an engine for growth and wealth creation is increasingly recognized by the development community. The increase of FDI flow to developing countries, from \$37 billion in 1990 to almost \$500 billion in 2005, is encouraging bilateral and multilateral agencies to adjust their aid policies to better facilitate FDI and its flow to needy countries. The marketplace is a new creative way to achieve social goals.

Connecting the Emerging World

More than 50 % of the world population has never made a phone call. In developing markets, high-speed and low-cost bandwidth, which the developed world can take for granted, is still almost non-existent. One can obtain 64–128 kbps for reasonable sums of money in developed markets. It is as though the bandwidth revolution has bypassed the developing markets even as the LANs have kept pace with the rest of the world. This is not to say that things are not changing: DSL, fibre, fixed wireless and cable solutions are promising to usher in broadband, but the impact of these initiatives, including wireless broadband access, is not as widespread as desired.

Historically, the solutions were proprietary technologies that operated in a few limited spectrum bands (e.g. LMDS, MMDS). These solutions were also limited to 'line-of-site' deployments (i.e. relatively flat, structured terrain where the home receiver required a direct line-of-site to the base station tower).

More recent BWA solutions, along with the WiMAX standard, are overcoming these initial technical limitations and expanding the addressable market. WiMAX will help BWA become a more mature, robust solution for a broader customer base. WiMAX is intended to improve BWA capabilities so that it can provide reliable voice, data and video services across wider operating environments.

WiMAX has been billed as an affordable way to bring the high-speed Internet to poorer and rural regions around the world and cover entire countries with seamless high-speed Internet access for viewing video, making phone calls and completing other data-intensive tasks. Further, WiMAX can meet the expected market demands for basic telephony in developing countries.

In the near future, the demand for WiMAX infrastructure will outpace the demand for 3G/UMTS infrastructure in places such as Latin America, Russia and India. In markets such as these, the teledensity (i.e. the ratio of phone lines to people) remains in the 10–40 % range and their governments are investing in communications infrastructure to help drive economic growth, enabling phone calls and other data-intensive tasks.

7.3   MARKET SEGMENT

Enterprise and Commercial Businesses

Enterprises and commercial businesses are increasingly seeking customizable services from their service providers. They not only expect higher bandwidth connectivity services to support demanding applications but also want the bandwidth delivered at greater levels of granularity. Their need for tailored bandwidth encompasses much more than simply data. Service providers need to provide support for data, voice and video traffic according to their customers' specific business objectives and applications.

Traffic is also no longer treated in the same fashion. For example, terminal traffic should be latency-optimized, while voice or video traffic should experience minimal loss and no packet reordering. At the same time, not being able to cost-effectively connect multiple offices and branches leads to delayed relay of information, which in turn impacts decision-making. In today's world, the mantra needs to be 'real-time'. In the case of emerging markets, this can be modified to 'near real-time' (or as close to real-time as one can get given the infrastructure). Reducing the time delay for information becomes the key challenge.

The new service model is becoming one of 'mass customization' rather than 'mass production.' Providers that can customize services for individual customers are more likely to realize the maximum revenue with these customers – all their requirements are acknowledged and fulfilled. Providers and enterprise users are understandably enthused by the potential of WiMAX to alleviate this difficulty. The emerging broadband wireless technology gives users more room to roam. It offers flexibility, ease of use and built-in security.

WiMAX can be an enterprise's wireless everything with potential to deliver 1.5 mbps/s or better bandwidth for wireless applications. The products can replace or supplement existing high-speed connectivity to administrative offices at a fraction of the cost of fixed T1 lines. WiMAX can also provide physical route diversity that goes beyond leasing a fibre-optic line for redundancy purposes from a secondary carrier.

Regardless of whether enterprise users opt to build networks or buy them as managed services, WiMAX is poised to meet the promise of broadband wireless that cellular, microwave and satellite technologies cannot.

Small and Medium-sized Businesses

This market segment is very often underserved in areas other than the highly competitive urban environments. The WiMAX technology can cost-effectively meet the requirements of small- and medium-sized businesses in low-density environments and can also provide a cost-effective alternative in urban areas competing with DSL and leased line services.

Besides being a good fit for many small- and medium-sized organizations, it turns out that this emerging technology has another major benefit: it will bring fast access to many businesses, particularly in smaller markets that do not have many connectivity options.

Municipalities and Local Government Bodies

Municipalities and local government bodies have historically made investments in essential services and infrastructure to improve the quality of life and/or increase economic development for their community. Local governments commonly pave streets, supply water and gas, remove rubbish and provide electricity.

Recently, many cities have also invested in communications and information services and infrastructure. There is a good argument for city/municipal ownership of all critical utilities as a way to enhance the reliability and security of critical infrastructure. As a user, regulator, economic developer and the community's infrastructure provider of last resort, cities are intimately involved with the local communications infrastructure, yet have very little input or much-needed control.

As the debate heats up regarding homeland security in the USA and public universal service across the globe, many local governments must and will take an active role in the communications services development.

Residential and SoHo

Today this market segment is primarily dependent on the availability of DSL or cable. In some areas the available services may not meet customer expectations for performance or reliability and/or are too expensive. In many rural areas residential customers are limited to low-speed dial-up services. In developing countries there are many regions with no available means for internet access. The analysis will show that the WiMAX technology will enable an operator to economically address this market segment and have a winning business case under a variety of demographic conditions.

7.4 MARKET STRUCTURE

Demographics play a key role in determining the business viability of any telecommunications network. Traditionally, demographic regions are divided into urban, suburban and rural areas.

Urban

Urban areas are considered to be the main market, rightly so, as urban markets provide the majority of business to most telecom operators. Broadband access is widely available; wireless broadband is also available but is costly. The situation varies widely depending upon status of development. Generally cable or DSL are available universally. Other characteristics of an urban market are:

- centrally located;
- high residential population density with more spending power than the national average and the highest spending power in comparison to other markets in the country;
- well spread infrastructure with the highest quality available in the country;
- highest density of business establishments and the centre of the majority of business activities.

Suburban

Suburban areas are considered an add-on to urban markets, as suburban markets are often seen as an extension of cities and are served by the infrastructure available in the city, with little modification by most telecom operators. Broadband access is available but not reliable and lacks quality; cable and DSL are not available universally. Other characteristics of a suburban market are:

- the distance from major metro areas is not very substantial;
- low-to-moderate residential population density with more spending power than the national average, but lower in comparison to the urban population;
- substantial infrastructure in comparison to rural communities but lacks quality and reliability, especially the telecom infrastructure;
- quite a few business establishments but mostly residential.

Rural

Rural areas are considered the final frontier of the telecom universe as rural markets are complex and often mysterious to telecom operators.

Broadband access is very sparse, if any; cable or DSL (relying on dial-up or satellite) so far only serves rural areas in a few privileged nations. Other characteristics of a rural market are:

- the distance from major metro areas is quite substantial;
- low residential population density with substantially lower spending power than the national average and in comparison to the urban population;
- very little or nonexistent infrastructure, especially telecom infrastructure, which is at minimal levels even in the overdeveloped world;
- little business establishment, agriculture and related activities being the main activities in most rural communities.

8

Economics of WiMAX

Providing cost-effective, affordable wireless bandwidth (almost) every-where is one of the key success factors for future wireless systems. As the success of the Internet is largely attributed to the fact that it is virtually free of (incremental) charges, it is generally perceived that wireless data communications have to provide services in a similar way. At the end of the day, WiMAX is all about delivering low-cost wireless broadband access (WBA).

8.1 WHY WIRELESS?

While certain drawbacks of wireless technology do exist, there are quite a few benefits that make the implementation of wireless solutions very attractive. The following list demonstrates the primary and most sub-stantial benefits of using wireless technologies:

- Wireless incorporation offers an all-inclusive access technology collection to operate with existing dial-up, cable and DSL techno-logies.
- The nature of wireless is that it does not require wires or lines to accommodate the data/voice/video pipeline; the system will carry information across geographical areas that are challenging in terms of distance, cost, access and/or time.
- While paying fees for access to elevated areas such as towers and buildings is not atypical, these fees, associated logistics and

The Business of WiMAX   Deepak Pareek
© 2006 John Wiley & Sons, Ltd

contractual agreements are often minimal compared with the infrastructural costs of other broadband access technologies.

- Businesses can generate revenue in less time through the deployment of wireless solutions because a wireless system can be assembled and brought online in as little as 2–3 h.
- This technology enables service providers to sell access without having to wait for current providers to provide access or backhaul.
- Wireless technologies play a key role in quickly and reliably extending the reach of cable, fibre and DSL markets, while also providing a competitive alternative to broadband wireline.

8.2   WHEN IS WIRELESS RIGHT?

Today WBA already seems to represent an economically viable alternative in rural areas, which often are underserved in terms of broadband access. An xDSL cable-based deployment will have significantly higher capital costs than a WBA deployment. xDSL capital costs can range from $100k to millions of dollars while the cost of a BAW base station ranges from $10k to $100k. The economics generally become better where the wired infrastructure is poor and the geographic region is more dispersed.

So far the incumbent network operators have only covered local exchanges (LE) over a given size (in terms of number of telephone lines) with DSL from economical criteria. LEs with few lines – under the threshold specified by the incumbent – and/or with too expensive backhaul costs are left out of the DSL coverage plans. The dramatic cost reduction of DSL equipment during recent years and the development of smaller units, e.g. 'mini' or 'micro' DSLAMs, have steadily brought this threshold down and are responsible for explosive growth of DSL over the past few years, but backhaul costs are still prohibitive in many areas.

Another key issue with DSL is that in many cases, even for areas with DSL coverage, subscribers cannot be served even though they are close to the central office. In such cases either the wireline to the subscriber is not economically viable or the increase in subscriber numbers is not economically feasible for that LE. Except some parts of Europe and North America, DSL does not serve even 75 % of the subscribers falling under DSL coverage. For the remaining subscribers wireless solutions are necessary. There are two segments of the 'broadband residual market' or 'unserved DSL areas' in DSL coverage:

- prospects in rural and remote areas with few telephone lines and/or high backhaul costs;
- prospects with too long or poor quality copper loops – these occur in all COs – and where CO capacity is an issue.

Sometimes a typical 0.5 Mbps DSL access will not be adequate for new services. Owing to the shorter reach of DSL for higher capacities, the 'residual market' with respect to these high-capacity levels will increase.

Other drivers to use wireless are low deployment costs and time as well as low recurring costs. Also, the ageing existing copper plant and difficult terrains make wireless an attractive option. A larger footprint (coverage) of different wireless technologies is therefore foreseen to provide high-capacity services to both fixed and nomadic users.

Traditionally, a profitable business case for wireless technology can be built in a range of demographic environments. However, underpinning the model are high-population-density, steep adoption curves and high ARPU assumptions. When the population density drops, the viability of network deployment can quickly fall away.

In addition, the marketing costs required to compete in a developing market can significantly impact project profitability. Without significant expenditure, the ability of a wireless broadband operator to win a share from existing operators should not be assumed. Assessing the viability of wireless broadband deployment requires a detailed and integrated assessment of market dynamics.

Using a 5 year investment horizon, networking a typical metropolitan area of 7000 square kilometres does not become an NPV-positive business proposition until the population density is greater than 135 potential subscribers per square kilometre. However, with the advent of WiMAX, this situation will change completely.

8.3 THE ECONOMIC ANALYSIS

One of the theoretical business cases presented by one of the many market research companies is built on the idea that a carrier might build a WiMAX network that covers 75 % of the US population at a cost of

$1.5 billion covering 85 million potential subscribers. The theoretical case relies on a $45-per-month charge and results in the carrier's breaking even at 3.35 % penetration, or 2.87 million homes.

WiMAX provides more a viable business case with wider profitability zone because of the low cost and the flexibility to cherry-pick subscribers along with possibility of tiered service delivery.

A WiMAX base station will have typically between four and eight sectors (channels), usually in a licensed band and capable of supporting up to 1000 registered CPEs per sector. Taking six-sector base stations with 100 subscribers per sector (600 per base station), with a 10:1 bandwidth oversubscription factor, the system is capable of downlink bursts of up to 14 Mbps and uplink bursts of up to 4.7 Mbps in a 7 MHz channel, with a range of less than 5 km from base station.

If we extrapolate this to a 10 MHz channel, the service provider would be able to offer peak rate of 20 Mbps on downlink and 7 Mbps on uplink (assume 75:25 uplink:downlink utilization in TDD) or 40 Mbps on downlink and 14 Mbps on uplink for 20 MHz channel – with the freedom to provide granular control of the bandwidth each direction in 128 kbps increments owing to the power of the 802.16 MAC.

The Cost Game

The challenge of providing flat rate, wireless access at the cost of fixed Internet access is indeed hard. The conventional cellular concept does not scale in bandwidth in an economical sense. The cellular systems include both the radio access network and the core network components, which have different cost and capacity performance. The more decentralized WLANs have a slightly shifted radio vs core network performance relation due to short range and high access capacity.

Disruption Advantage

A disruptive technology will always have lower cost than incumbent technology. This theory is, however, difficult to measure and test in the emerging wireless market for several reasons.

The first is the necessity to define, delimit and compare the relevant parts of the networks and technologies as they are built and configured differently:

- front end network – base stations;
- backbone infrastructure;
- switches and service platforms.

Secondly, as we are discussing networks that both historically and in the future will be capital-intensive, a cost comparison also has to be based on equal principles of capitalization. If the levels of expensing vs capitalization, OPEX vs CAPEX, are very different, the short-term cost picture will also be very different.

Thirdly, when it comes to end-user equipment/CPE, this is also a challenge. An intuitive starting point here might be to compare the cost of cellphones with portable PCs and PDAs. The problem is that a portable PC contains a large variety of functionality that is network-independent and irrelevant for a cellphone. It is also likely that the consumers already have or will buy a portable PC/PDA for other reasons than for mobile network services.

Thus, the presence and penetration of end-user equipment will be very high at the point in time where the convergent technologies and networks are launched.

Implications of Standardization

The lack of standardization is one of the major challenges for the present BAW technologies. This is especially relevant for point-to-multipoint radio technologies. Within the Wi-Fi area (802.11x) the technology has been standardized with a price/performance curve, as we know from the PC-component industry. The standardization work related to WiMAX is expected to create a similar price/performance curve. WiMAX is based on the 802.16e (OFDMA) industry standard and can be implemented without the costly, proprietary interfaces and royalties found in 3G networks.

The development of WiMAX industry standards (802.16) will significantly improve the economics for both network equipment and CPE. Major manufacturers such as Intel, Motorola, Fujitsu, Siemens and Alcatel are already committed to development in accordance with these new standards. Their focus is on bringing scale to the market.

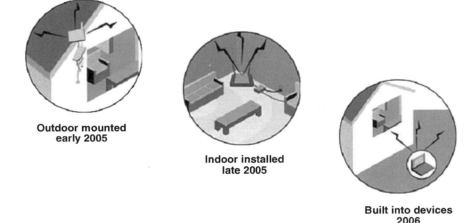

**Outdoor mounted
early 2005**

**Indoor installed
late 2005**

**Built into devices
2006**

Figure 8.1   WiMAX CPE evolution

Nearly all laptops and a large number of PDAs and cellphones are already Wi-Fi-enabled. It is expected that chipset manufacturers such as Intel will target standardizing and embedding WiMAX chipsets in laptops and other mobile devices within 2006.

The manufacturers of BAW equipment are also always interested in less expensive chipsets. A standardization of BAW technologies will result in interoperability, which in turn will bring plug-and-play products. That should imply that in years to come WiMAX vendors will no longer have to provide end-to-end solutions in a complete network (Figure 8.1). They will be able to specialize in different components such as base stations or wireless modems. Such specialization will result in competitive pricing and value-added innovations. The standardization of the BAW technologies will also most likely provide easier upgrade paths to future technologies, without the costly need to dispatch technicians or physically run wires.

8.4   THE BUSINESS CASE FOR WiMAX

As there are many different levels on which to study business cases, from a political to a social level, it is often difficult to decide where to stop the analysis. In the cases stated here, a purely techno-economic plane has been employed to understand the processes and partnerships that take place between different business entities in their pursuit of delivering a product.

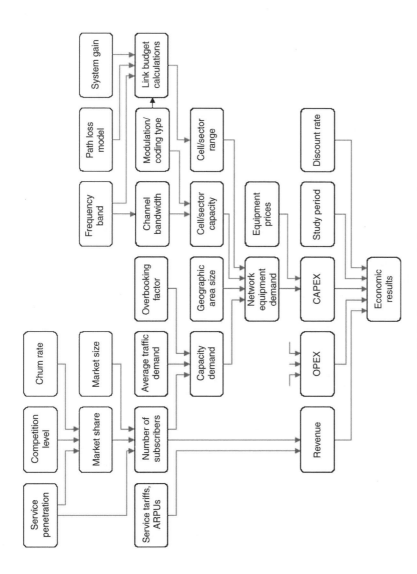

Figure 8.2 Techno-economic framework

Techno-economic Framework

Broadband services open up new revenue possibilities for operators. At the same time, the upgrading of access networks to support these services requires large investments, and many alternative technologies can be used. Techno-economic analyses are required to discover the optimal technologies and systems for different environments. We will use these analyses to find out whether or not the WiMAX networks are really competitive with the other technologies (Figure 8.2).

For the purposes of this section 'techno-economic analysis' is defined as an analysis seeking to determine the economic feasibility of a technology. Much of the terminology, methodology and tools related to these techno-economic analyses of broadband access networks have been developed in the various research projects funded by various bodies across the globe, especially the European Union, including those in the Table 8.1 below.

The techno-economic framework basically consists of the following building blocks:

- area definition – geography and existing network infrastructure situation;
- service definitions for each user segment with adoption rates and tariffs;
- network dimensioning rules and cost trends of relevant network equipment;
- cost models for investments (CAPEX) and operation costs (OPEX);
- discounted cash flow model;
- output metrics based on NPV.

Table 8.1   Research related to techno-economics of broadband networks

Project name	Research programme (framework programme)	Timeframe
TITAN (tool for introduction scenario and techno-economic evaluation of access network)	RACE II (FP3)	1990–1994
OPTIMUM (optimized architectures for multimedia networks and services)	ACTS (FP4)	1994–1998
TERA (techno-economic results from ACTS)	ACTS (FP4)	1994–1998
TONIC (techno-economics of IP optimized networks and services)	IST (FP5)	1998–2002

Scenarios as Inputs

A number of choices, assumptions and predictions have to be made before proceeding to the techno-economic analysis of a broadband access network. These include the selection of the geographical areas and customer segments to be served, the services to be provided, and the technology to be used for providing the services. Assumptions and predictions are needed, e.g. on the level of competition in the market, the penetration rates of different throughput classes, and the price evolution of service tariffs and network components.

In the present framework, scenarios are used to depict the access network evolution from the existing situation to the long-term company target. A scenario is defined as a description of a network environment, including one or several operators providing a set of services to a number of users within a certain area and timeframe. A complete scenario is composed of the regulatory, environmental, service, and technology scenarios, each characterized by a number of scenario attributes.

The regulatory scenario describes the tariff structures and the revenues of the operators, as well as the sharing of the potential market between the operators. The attributes to be defined include the number of competitors both in the service market and access network provisioning, and the percentage shares of the competitors in these markets.

The environmental scenario describes the geographic and demographic characteristics of the area that is to be provided with a new network or a network upgrade. In addition, the existing network infrastructure is described.

The service scenario describes the services provided to the end-users by the service operators. The time evolution of the penetrations and the tariffs of these services are also defined.

The technology scenario describes the technologies, systems and architectures that are used to provide the selected services to the end-users. The evolutionary steps between the existing network and the final network architecture are defined. Also, the cost of network equipment and installation, together with the cost of operations, administration and maintenance (OA&M) procedures, is defined.

The scenario attributes are not fixed figures, but change over time. Accordingly, the attributes have to be defined as time series for the whole study period.

Expenditure

The costs of building and operating a broadband network can be divided into capital expenditures and operational expenditures. CAPEX includes the investments to the network infrastructure and devices, as well as the hardware required for the OAM&P functions such as network management systems and billing and charging systems. OPEX includes the labour costs and expenses originating from operating and managing the networks as well as costs related to, for example, marketing, sales and customer care.

Capital expense (CapEx)

CAPEX costs relate to investment in equipment and the design and implementation of the network infrastructure, e.g. site acquisition, civil works, power, antenna system and transmission. The equipment includes the base stations, the radio controllers and all the core network equipment. An example of CAPEX and the relationship between different types of implementation costs is shown below.

Components

- base stations;
- site preparation;
- service platforms;
- spectrum.

Operational expense (OpEx)

The operational expenditures of network operators are often referred to as OA&M or OAM&P costs, the letters representing operations, administration, maintenance and provisioning. According to the ITU Recommendation M.60:

- operations include the operation of support centres/systems as well as personnel and training required to install and maintain the network elements;
- administration ensures the service level once the network elements have established the service;
- maintenance includes carrying out the preventive measurements and locating and clearing faults; and
- provisioning makes the service available by installing and setting up the network elements.

The operational expenditures related to a certain project are often more difficult to predict than the capital expenditures. This is especially true when new network technologies are considered, as previous experiences or data are not available. Often, the OAM&P costs are simply derived from the capital costs using a proper coefficient.

OPEX is made up of three different kinds of costs:

- customer-driven, i.e. costs to obtain customer, terminal subsidies and dealer commissions;
- revenue-driven, i.e. costs to persuade a subscriber to use the services and network or costs related to the traffic generated, e.g. service development, marketing staff, sales promotion and interconnection;
- network-driven, i.e. costs associated with the operation of the network, e.g. transmission, site rentals, operation and maintenance.

The key factors are related to customer acquisition, marketing, customer care and interconnection. The fraction of OPEX to the overall cost changes over time; in the 'mature' phases OPEX is the vital factor. However, an estimate indicates that network-related OPEX is roughly 25–28 % of the total costs for the full life cycle.

Components

- Site leases
- Backhaul
- Network maintenance
- Customer acquisition

Profitability as Outputs

A prime result of a techno-economic analysis is whether or not the investment project in question is profitable or not. Commonly used measures to determine the profitability of a project include the project's net present value, internal rate of return and payback period.

Discounted cash flow model

When revenues, investments and all operational costs are estimated for each year during the period under study, the cash flow series CF (t) can be established:

$$CF\ (t) = Revenue\ (t) - Investment\ (t) - OPEX\ (t)$$

The time-value of money and risk is taken into account in the discount rate r. The discounted cash flow series DCF (t) is defined as

$$DCF(t) = CF(t)/(1 + r)t$$

The sum of all discounted cash flows produces one number called the net present value or NPV. The NPV is a measure of the value of a project. Put simply, if NPV > 0, the project is profitable, otherwise it is not.

The NPV of an investment project is the most favourable measure of profitability, and leads to better investment decisions than other criteria. The NPV of a project is calculated as the difference between the discounted value of the future incomes and the amount of the initial investment. The NPV rule states that a company should invest in any project with a positive NPV. The discount rate, also known as the opportunity cost of capital, represents the expected return that is forgone by investing in the project rather than in comparable financial securities.

The internal rate of return (IRR) of a project is closely related to the NPV. In fact, the discount rate that makes NPV = 0 is also the IRR of a project. The IRR rule states that a company should accept investment opportunities offering IRR in excess of their opportunity cost of capital. Although commonly used in many companies, the IRR has some pitfalls and deficiencies compared with the NPV method. The payback period of a project is the number of years before the cumulative incomes equal the initial investments. When using the payback rule in investment decisions, all projects that pay themselves back before a defined cut-off date are considered profitable. The payback rule has some major deficiencies, including the fact that it ignores all cash flows after the cut-off date. Furthermore, it does not take the time-value of money into account, but gives equal weight to all cash flows before the cut-off date.

Cash balance curve

The cash balance curve gives a simple and easily understandable overview of a project's profitability, and is a good tool to be used together with, for example, the NPV (Figure 8.3).

Risk and Sensitivity Analysis

As discussed earlier, the investment costs in access network upgrade projects are high. The lifetime of the investments is also expected to be

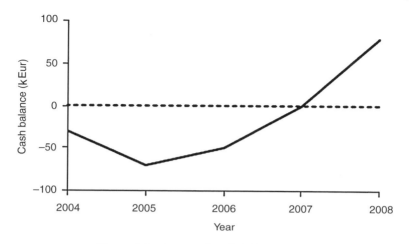

Figure 8.3 A typical cash balance curve

many years, requiring the operators to make predictions for the distant
future (Figure 8.4). These forecasts always hold a certain degree of
uncertainty, the main sources of which are the predicted service
demands, the competition between operators, the costs of network
components and the costs of operating the new network architectures.
These uncertainties and their effects on the viability of the investment
projects are assessed by means of risk and sensitivity analyses. A
common approach to handling risk in investment decision processes is

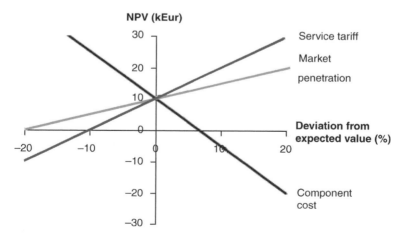

Figure 8.4 A typical sensitivity graph

sensitivity and scenario analyses. This approach aims to incorporate the project risk directly within the NPV formula.

Sensitivity analysis is a simple technique used to locate and assess the potential impact of risk on a project's value. The aim is to identify the impact of changes in key assumptions on the profitability (e.g. the NPV) of the project. The results of the sensitivity analysis can be plotted in the so-called sensitivity graph that offers an illustrative view of the sensitivity of the variables. The sensitivity of each variable is reflected by the slope of the line – the steeper the line, the greater the impact of changes on the NPV.

TERA Tool

The techno-economic analysis carried out here is done using a tool which is very similar to the proprietary tool named TERA. The TERA tool is a spreadsheet-based application for techno-economic assessment of communication networks and services. It was developed within the European Union ACTS (Advanced Communications Technologies and Services) Programme. TERA tool enables techno-economic evaluations and strategic analyses that combine high-level parameters such as density of subscribers and service penetration, with relevant low-level parameters such as costs of key network components. As such, the TERA tool is suitable for the present purpose, straightforward to use and adapts easily to the different scenarios. The outputs of the tool are easy to interpret and traceable to the inputs due to the visibility of the formulas in use. However, we will not go into a detailed analysis and will concentrate on the result of the analysis (Figure 8.5).

Applying the Analysis – WiMAX

Let us now apply methods of the previous section to analyse the economic aspects of WiMAX deployments in different kinds of environments. Here we try to answer the following broad questions:

- Is WiMAX a threat to established operators?
- Is WiMAX a possibility for established operators?
- What are the most attractive environments for WiMAX?
- Is WiMAX capable of becoming the technology of choice for new entrants?

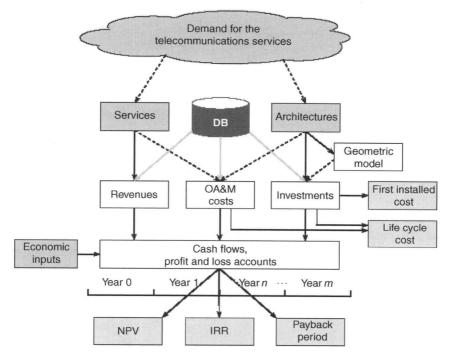

DB integrated cost database

Figure 8.5   TERA tool

- Could WiMAX provide broadband services more cost-efficiently than DSL?

The analysis is made for a study period of 5 years, 2004–2008. The analysis is made for six different types of areas or environmental scenarios:

- two urban areas, one from a developed and the other from a developing country;
- two suburban areas, one from a developed and the other from a developing country;
- two rural areas, one from a developed and the other from a developing country.

Each of the scenarios is characterized by their geographic areas and household densities. The data used as the base in forming the scenarios

Table 8.2 Attributes of environmental scenarios

Scenario	Geographic area ($L \times L = A$ km^2)	Household density (1/km^2)	Numer of households	Number of telephone exchanges
Urban area – developed	$2 \times 2 = 4$	5000	20000	2
Urban area – developing	$3.2 \times 3.2 = 10$	2000	20000	2
Suburban area – developed	$4 \times 4 = 16$	1000	14000	2
Suburban area – developing	$5 \times 5 = 25$	500	14000	2
Rural area – developed	$30 \times 30 = 900$	5	5000	16
Rural area – developing	$50 \times 50 = 2500$	2	5000	36

are hypothetical. The scenarios were deliberately constructed so that in each scenario the achievable number of subscribers would be roughly the same, also taking into account the expected market share evolution in each environment. In each case, the expected number of subscribers at the end of the study period is about 1500.

Results

The business case for WiMAX was analysed for several scenarios (Figures 8.6–8.11) – urban, suburban and rural areas – and from the

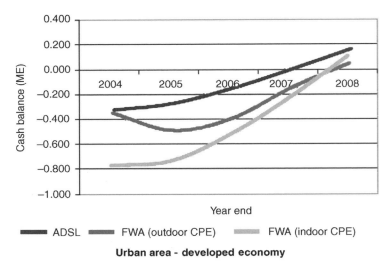

Figure 8.6 Cash balance graph scenario 1

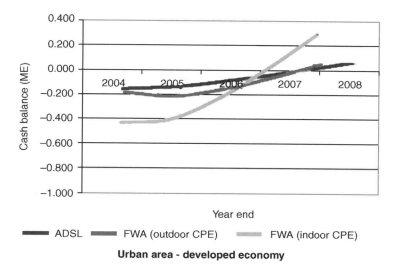

Figure 8.7   Cash balance graph scenario 2

point of view of incumbent and new entrant operators. The analysis was based on a number of assumptions such as cell sizes, deployment of competing technologies and subscriber density.

The most notable factor was the percentage of ADSL deployment in the WiMAX service area and, thus, the analysis was made for different

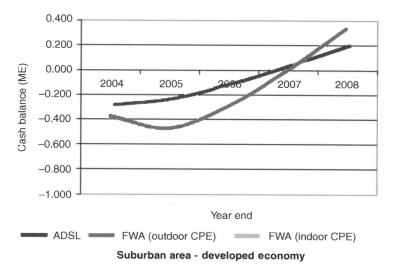

Figure 8.8   Cash balance graph scenario 3

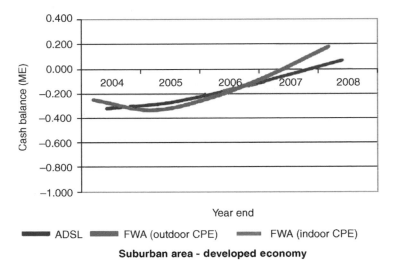

Figure 8.9 Cash balance graph scenario 4

values of this factor. The results are presented in the matrix below. The matrix shows the feasibility of WiMAX solutions in different scenarios considered for the study. The analysis shows the accumulated cash flow for urban and rural areas with high and low levels of ADSL penetration.

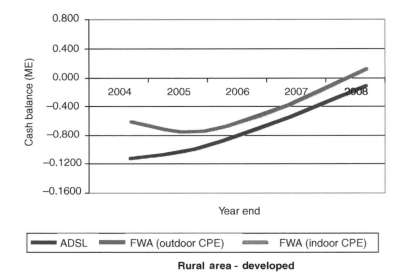

Figure 8.10 Cash balance graph scenario 5

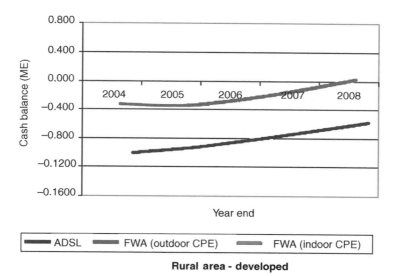

Rural area - developed

Figure 8.11 Cash balance graph scenario 6

WiMAX does not seem to be economically feasible in providing broadband Internet access to households, and in the urban areas of developed countries, the NPV figures of WiMAX project are negative. Also, in this case the payback time of the project is longer than the study period of 5 years.

Table 8.3 Matrix of opportunity – demographic

	Urban	Suburban	Rural	Ex-urban
Overdeveloped	Medium	High	High	High
	Potential	Potential	Potential	Potential
	High	High	Medium	Low
	Competition	Competition	Competition	Competition
Developed	Medium	High	High	High
	Potential	Potential	Potential	Potential
	High	Medium	Low	Low
	Competition	Competition	Competition	Competition
Underdeveloped	High	High	High	High
	Potential	Potential	Potential	Potential
	Low	Low	Low	Low
	Competition	Competition	Competition	Competition
Least developed	High	High	High	High
	Potential	Potential	Potential	Potential
	Low	Low	Low	Low
	Competition	Competition	Competition	Competition

Table 8.4   Matrix of opportunity – sector based

	Residential	SME	Enterprise	Public
Overdeveloped	Medium Potential High Competition	High Potential High Competition	Medium Potential High Competition	High Potential High Competition
Developed	Medium Potential High Competition	High Potential High Competition	High Potential High Competition	High Potential High Competition
Underdeveloped	High Potential Low Competition	High Potential Low Competition	High Potential Low Competition	High Potential Low Competition
Least developed	High Potential Low Competition	High Potential Low Competition	High Potential Low Competition	High Potential Low Competition

8.5   BUSINESS CONSIDERATIONS

The advances promised from WiMAX technologies represent a radical shift in the wireless access business model for the scores of manufacturers and carriers involved. However, it is also important to note that the shifting is far from over.

WiMAX Forum Certified products are not yet available. The first product certifications will not be issued before 2005. In order to ensure that a large number of vendors reach certification immediately, the first WiMAX profiles still include many different optional features, such as diversity, space–time coding and ARQ.

WiMAX will be deployed in multiple phases. Volume will be driven by portability/mobility applications, such as notebooks and PDAs. These volume benefits, as well as forward-compatibility of the evolving standard, will only be addressed in the 802.16e standard, scheduled for release in 2005.

The Significance of WiMAX

WiMAX will be used in urban, suburban and rural areas, particularly where other broadband means are not available or installations are expensive. Competition with DSL will not be fierce in areas where DSL

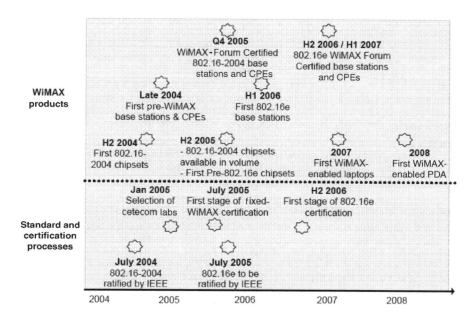

Figure 8.12 WiMAX availability

is already established due to relatively low DSL costs and high penetration. Furthermore, a high density of WiMAX base stations will be needed in urban and suburban areas to be able to serve customers with self-installable CPE and reasonable data rates. In fact, the cell sizes under those conditions are only a few hundred metres (Figure 8.12).

WiMAX is likely to play an important role in serving rural areas. There, cell sizes of 5–10 km are possible, requiring outdoor antennas at the customer premises.

Can WiMAX Compete with DSL?

Competition is tough . . .

- DSL providers see their infrastructure upgrades as sunk costs.
- DSL is often bundled with fixed services.
- DSL providers are usually not yet profitable, but it is possible to gain market share if:
  - o fixed coverage is limited, or service is inadequate;
  - o fixed service offering is expensive, has limited features, or does not allow flexibility;

○ the WiMAX service provider can offer a richer service that
   includes VoIP, portability, mobility or other advanced features;
○ the WiMAX service provider charges comparable (or lower)
   prices.

Strengths, Weaknesses, Opportunities and Threats (SWOT)

Strengths

Based on proved OFDM techniques (inherent robustness against multi-path fading and narrow band interference), the strenghts are:

- low cost of deployment and operation (~$33 per home vs
  $300–600 for DSL);
- high speed (75 Mbps) and long range (50 km);
- adaptable and self-configurable system;
- centralized control in MAC enables simultaneous, varied QoS
  flows.

Weaknesses

In addition to the general disadvantages for wireless broadband, specific shortcomings exist for different wireless standards including WiMAX. Some of them are:

- currently high power consumtion (still far from penetrating
  portable mobile devices);
- mobility not yet fully specified – could become complex to implement.

Higher frequency WiMAX signals, around 10 MHz and above, take on line-of-sight characteristics that require a free path for signal transmission.

The standard is not backwards-compatible with current wireless technologies; in other words, current wireless equipment of the 802.11 standard cannot directly receive or transmit 802.16 signals. As a result, completely new wireless equipment needs to be purchased by each WiMAX subscriber.

A final weakness for the WiMAX standard pertains to the location of the customer premises equipment. These subscriber station antennas can be either indoors or outdoors for signal reception. However, indoor antennas do not guarantee signal reception in every building. In fact, a WiMAX infrastructure with any indoor antenna would require smaller cell sizes and result in higher costs for the network. The need to install outdoor antennas may turn users away from WiMAX networks.

Opportunities

- High-speed wireless infrastructure.
- Cellular infrastructure for converged networks.
- Last-mile solution for broadband wireless access.
- CAPEX of ~$3.7 billion needed to penetrate 97.2 % of USA homes (InStat/MDR).

Threats

- DSL/ADSL technologies widely deployed.
- Cellular penetration is very high and growing.
- Possible wide deployment of 3G.
- Widespread success of 802.20 standards.

Risk

There are many risks and uncertainties which confront new telecommunications technology development and its adoption. Not all telecommunications risks are recognized and addressed; however, the major risks and uncertainties are considered. The risks are categorized into four different sections: technological uncertainties, demand uncertainties, operational uncertainties and legal and regulatory uncertainties.

Technological uncertainties

The 'digital divide' is one of the key risk factors. It is important for operators to consider the risks of furthering the digital divide, as it represents potential high-level changes to the regulatory structures and perhaps service obligations which may not be economically attractive. Convergence is the second technology-related risk. Convergence is a supply-driven phenomenon; this change is the outcome of technological innovation. However, if the market is economically restricted, the potential for some features of convergence to be widespread must be examined.

Demand uncertainties

Demand uncertainties involve the risks and doubts concerning customer reaction to a product or service and its acceptance. As demand is vital for success of any new technology, this form of uncertainty is also a key risk factor.

Market uncertainty

Most risks associated with new or improved products originate from technological and/or market uncertainties. There is never certainty about the consumers' reaction to launching a new product, whether it is a radical or an incremental innovation. Technology-driven markets are more uncertain than purely demand-driven markets, since radical innovations are often based upon presumed or extrapolated needs, as opposed to existing, identified needs in demand-driven markets where the improvements are often incremental. Even where a great need for technology exists, the quality and dynamics of this demand are difficult to predict.

Legal and regulatory uncertainties

The success of attracting and keeping telecoms operators and service providers relies mainly on the stability and viability of the regulator. Should this environment be subject to political interference or mismanagement, the development and progress required of the ICT sector are unlikely to materialize. Licence obligations, while perhaps necessary to fulfil certain obligations made by government to the people, cannot be overly severe or inflexible.

Experience has shown that a flexible service obligation allows operators to fulfil obligations, achieve some profitability in rural areas and stimulate indirect benefits such as job creation and development of local industries. However, where licence obligations are subject to repeated change and the regulator is not independent or endowed with sufficient authority, the environment is likely to suffer. A second worry is corruption in the legal and regulatory authority.

Telecommunications disputes that require legal arbitration and judgement have the potential to severely impact the perception of the sector by investors, operators and businesses. This can significantly hamper growth.

Political, Economic, Social, Technical (PEST) Analysis

PEST analysis is important so that an organization can consider its environment before starting any new initiative. In fact, environmental analysis should be continuous and influence all aspects of planning. The organization's marketing environment is made up of:

- the internal environment, for example staff (or internal customers), office technology, wages and finance;
- the microenvironment, for example external customers, agents and distributors, suppliers, competitors;
- the macroenvironment, for example political (and legal), economic, socio-cultural and technological forces. These are known as PEST factors.

Political Forces

- Community.
- Commercial.
- Revenue.

Economic Forces

- Equipment.
- Installation.
- Backhaul.
- Interconnection to other networks.
- Network operations centre (status, failure alarms).
- Network maintenance (equipment repair, interference resolution, hackers, viruses).
- Customer support (access to the network, blacklisting users, help desk).
- Capacity/expansion planning.

Social Forces

- Demography – urban, suburban, rural.
- Development – underdeveloped or developed.

Technical Forces

- Radio Standards: Wi-Fi a/b/g, WiMAX, 4.9 GHz, etc.
- Interference issues: these bands are not protected.
- Quality of service: if network is shared, who and what has priority?

Total Cost of Ownership

The broadband wireless access market is expected to increase dramatically over the next 10 years as the demand for high-speed Internet access
explodes. However, currently, less than 10% of Americans have broadband; overseas, the percentage is even smaller.

When enterprises first started deploying wireless LANs, the cost
justification seemed straightforward enough, once prices came down to
a certain level. We still paid a premium for wireless LAN hardware and
software, of course, but that was offset by the wired LAN (W-LAN)
cabling and installation costs we avoided. Then there were all the soft-
dollar W-LAN benefits associated with the convenience of being able
to move around in a facility and stay connected.

Gartner pioneered the notion of a total cost of ownership (TCO) for
technology. TCO takes into account all kinds of costs besides the
equipment or software vendor's price. As Gartner has shown over and
over, capital costs are often only a small part of the total. The Gartner
TCO model takes into account four broad categories of cost: capital, IT
operations, administrative and user operational. For TCO analysis there
are direct expenses and indirect expenses to be considered. Direct
expenses associated with acquiring and deploying a mobile wireless
solution may include the following:

- hardware – mobile devices, PC cards, SIM cards, extra batteries;
- software – hosted applications, middleware, security;
- services – design, integration, configuration, deployment, training;
- operations – airtime fees, technical support, user help desk, con-
  sumables;

- maintenance – software and hardware maintenance, repair, extended warranties, spares;
- building costs for wireless infrastructure as appropriate.

Indirect expenses for sustaining operations may include the following:

- downtime – back-up, restore, failures, lost work due to non-working conditions (no coverage, etc.);
- IT support – troubleshooting, technical help desk, testing, logistics, change management.

Other indirect expenses, or savings, that may be more difficult to quantify include productivity losses/gains, impact of a faster/slower response rate, user retention rates, ease of configuration, ease of updates (software and firmware), ease of migration and end-to-end quality of service rates. The above expenses, from an accounting standpoint, can also be categorized as the cost of equipment (CAPEX), the cost of equipment installation (OPEX) and the cost of maintenance (also OPEX).

Existing technologies – DSL, cable and fixed wireless – are plagued by expensive installations, problems with loop lengths, upstream upgrade issues, line-of-sight restrictions and poor scalability.

WiMAX answers these challenges by offering a scalable, wide-area wireless broadband access network that leverages unique smart antenna capabilities, requires zero installation at the end-user site and offers true non-line-of-sight operation, with the added benefit of portability.

Further, WiMAX provides telecommunications carriers with the lowest total cost of ownership – up to 50 % lower than DSL and cable and up to 70 % lower than fixed wireless technologies.

WiMAX is an all-IP, all packet technology with no legacy circuit telephony, which makes the operational expenses very low, thanks to the transport efficiency of IP for short bursty traffic such as data connection and single direction traffic such as voice.

The use of all-IP means that a common network core can be used, without the need to maintain both packet and circuit core networks, with all the overhead that goes with it. A further benefit of all-IP is that it places the network on the performance growth curve of general purpose processors and computing devices, often termed 'Moore's law'.

Computer equipment advances much faster than telecommunications equipment because general purpose hardware is not limited to

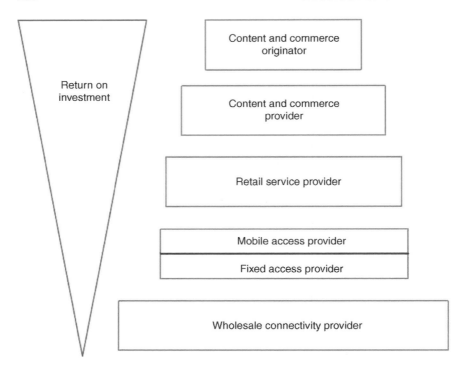

Figure 8.13   ROI for different service providers

telecommunications equipment cycles, which tend to be long and cumbersome. The end result is a network that continually performs at even higher capital and operational efficiency and takes advantage of third party development from the Internet community (Figure 8.13).

Return on Investment

Return on investment (ROI) analysis has served as a key decision-making tool for many organizations in new technologies, including WiMAX. The ROI for this new wireless broadband technology is significantly better than for DSL and cable as well as 3G.

The maths is not difficult. First, DSL/cable means truck roll for installation, which means a lot of investment in copper, trenching and finally the CPE. These difficulties are not faced with WiMAX, and also the cost of WiMAX installation is much lower than that of DSL. Second the installation of DSL takes weeks, while that of WiMAX takes hours, meaning a shorter gestation period, which is the period after investment and before revenue is earned. Finally, the maintenance of DSL/cable is an expensive and time-consuming activity, which is not the case with WiMAX, with its lower operation cost.

The start of 3G itself is shaky. Huge investments in terms of licenses make 3G an expensive proposition. As an example, in Singapore the highest amount paid in the BAW Auction was $2.27 million – Pacific Internet's bid for five lots of spectrum – compared with the $100 million that cellular operators forked out for their 3G licences in 2001. Thus 3G involves total revamp of infrastructure, which means huge upfront investment as well as the wasting of existing infrastructure. WiMAX gives an option of providing 3G services without extensive investment as existing infrastructure can be used with extension or the addition of WiMAX infrastructure.

For these reasons WiMAX provides better ROI than DSL, cable and 3G.

8.6   WiMAX BUSINESS MODELS

Enterprises, Internet service providers and mobile network operators are all looking for cost-effective ways to move voice and data amongst multiple, separate locations at broadband speeds. Copper and fibre optics solutions often fall short due to up-front costs, recurring leases from telecommunications companies and a lack of flexibility to scale with the operating organization. Broadband wireless has emerged as a means to fill these gaps and to provide a lower total cost of ownership than wired solutions, while maintaining or exceeding the reliability and performance of those technologies.

In recent years, we have seen many changes in the broadband wireless industry. While technology was evolving, evolution of the business models used in the industry followed. As services changed from voice-centric to data-centric services, operators had to adopt different business models.

Business models have to evolve and change with changes in technology as old business models will no longer work for industry players. The link between technology and business is tightly coupled. New technology will not work if there is no business case for it and if industry members do not support it, and industry members will not support a technology if they feel that it would not contribute to new revenue opportunities. Therefore, when technology evolves to a higher plane, business models describing processes and flows between the companies involved in the supply of the end product will also evolve to a different plane to take into account what the new technology and the services that go with it can offer.

As technology has evolved and transmission rates have become faster and include data services, changes have also occurred with the business models. The interrelationships between the different companies that work together to provide services to users have undergone changes as technology and the possibilities that came with new technologies have themselves changed.

Business models are very popular and used frequently to describe the interrelationships between firms engaged in producing a particular product. The concept of a business model is often confused with that of a business plan and is not really understood.

A business model is 'the organization of product, service and information flows, and the sources of revenue and benefits for suppliers and customers' or in other words 'the way a network of companies intends to create and capture value from the employment of technological opportunities.'

Generally, it can be said to be a representation or a description of how a group of companies interact with each other in order to produce product value for customers. In this section, we will explore the WiMAX business model.

Debating WiMAX

In recent years, the use of BWA solutions to solve connectivity needs for both commercial and residential applications has surged in popularity. Overall system costs dropped dramatically from the initial BWA deployments made using multichannel multipoint distribution service (MMDS) technology in the 1990s to those using pre-WiMAX solutions now, and will drop again in the future as certified WiMAX systems make their appearance and gain volume.

The use of licence-exempt radio frequency spectrum has enabled thousands of service providers to offer competitive services to leased lines, ISDN, DSL and cable modems. Today, the BWA market is shifting from initial small-scale installations to large-scale regional and national deployments as confidence in the technologies and the ability to achieve profitable business models accelerates. BWA has been proven to be reliable technology with deployments around the world.

BWA successfully serves a variety of market and application needs today from last-mile access to video surveillance to metropolitan area networks.

WiMAX represents the next evolution in broadband wireless technology and is backed by Intel. There is significant debate regarding the business model for WiMAX. Originally WiMAX was intended to be an alternative or substitute for DSL and cable modem broadband, particularly where neither technology is available. However, this model does not offer many opportunities in the developing world, including North America and Western Europe, where broadband coverage is nearly total or very extensive and there are many retailers of xDSL vying for the customer's attention.

There is, of course, the possibility of using WiMAX to fill the gaps in areas where wireline broadband access is not feasible or for an operator looking to provide something rather different such as city-wide solutions to the public sector. Nonetheless, to compete head-on with DSL would really mean offering the same service using a different technology.

WiMAX can be deployed profitably in many ways and with various business models, but there are typically three broad business models which are central to this technology: last mile, backhaul and metro access network models are the basic premise from which various other business models are derived.

Last mile access – with a personal touch

As total Internet access subscriber rates rise by leaps and bounds each year, service providers of all sizes and kinds are facing a common challenge: differentiation of service offering in a competitive market. Further, providers must address customer demand for quick service delivery and high-speed network access at a low cost. The price associated with customer acquisition costs and ongoing last-mile maintenance puts pressure on service providers to find new ways to cost-effectively deploy differentiated Internet access services to large numbers of customers.

WiMAX gives wireless ISPs and DSL/cable providers today's fastest pathway to new markets and revenue. Whether well established and looking to expand or smaller and newly established, service providers of all types can immediately and cost-effectively create networks or reach out from established points-of-presence to capture new customers.

Without the delays and costs of leasing or building a wired infrastructure, BWA networks allow secure and reliable access to high-speed data, voice and video services. Internet access can be extended to business parks, apartment complexes, school districts and even rural communities several miles/kilometers away – all in a matter of days.

Further WiMAX operators are going to offer fixed competition to drive asset utilization up to the necessary levels, but the new opportunity is to provide personal broadband, or, in other words, 'broadband that goes with you' (Figure 8.14).

Personal broadband

Personal broadband does not necessarily require the kind of truly mobile broadband connection that would support, for instance, an Internet connection with a fast (multi-Mbps) throughput while the user is travelling on a high-speed train. For such personal broadband use,

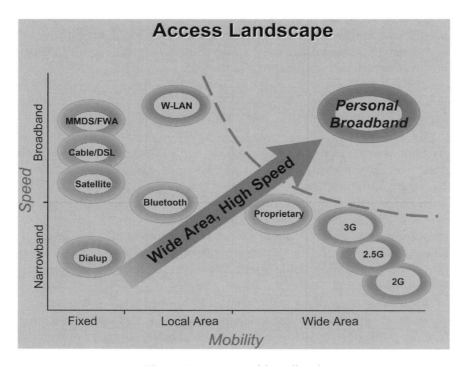

Figure 8.14 Personal broadband

urban commercial users (the majority of the market) will mainly require in-building broadband coverage and a portable connection that is available to a user while walking or in a slow-moving vehicle.

There are well-known market niches for personal broadband, particularly amongst small business owners and, of course, 'road warriors'. However, while these are well-understood and attractive niche markets (where WiMAX should have clear advantages of speed and simplicity over 3 G technologies), ultimately the most interesting area will be in the new markets opened up by wide area wireless broadband. Putting the 'ether' back into Ethernet by restoring it to a wireless connection is likely to create unpredictable changes in the ways in which broadband is used.

Just as the mobile phone now generates more voice revenue than the fixed network in many countries, would it not also be possible, at least eventually, for wireless broadband revenues to exceed those of wireline broadband, if it were simple and cheap enough?

For example, for in-vehicle entertainment, flexible CCTV and security systems, WiMAX devices could represent a user's second or third broadband connection. This situation is entirely feasible – one should remember the time when nobody believed that mobile penetration would exceed 100 %!

Similarly, personal broadband could provide a facility for voice over wireless broadband that would act as a pseudo-mobile service, which could prove attractive for certain market segments. Personal broadband could prove disruptive to these and other existing markets in the long term.

Ethernet economics – the general and rapid improvement in the cost performance of Ethernet devices and systems – has often surprised the telecoms industry, and WiMAX looks set to provide the next example over the next few years.

Backhaul

Delivery of a broadband service to customers depends not only on the broadband access network, the so-called 'last mile', but also on a means of connection to the mainline or backbone networks that forms part of national and international data transmission networks. This connection is known as backhaul and has been called the 'middle mile'. Backhaul is a significant issue, since high-capacity networks are normally found in large towns, and obtaining connection to them is a substantial factor in the cost of rural broadband services. Backhaul to the nearest available main network node can be addressed by a variety of technologies:

- optical fibre;
- DSL-family technologies over copper circuits;
- radio links;
- satellite links.

Secondary backhaul is necessary for hybrid broadband access solutions. This occurs where for reasons such as cost or technology limitations the broadband access ('last mile') is provided from a point or points remote from the local exchange building (or point of presence). This creates a hybrid technical solution using one method of broadband access from the remote distribution point but relying on another for secondary backhaul between the local exchange and remote distribution point.

WiMAX provides the most cost-effective solution as backhaul to Wi-Fi and cellular networks.

Metropolitan Area Networks

The new generation of MAN-sized wireless technologies based on IEEE 802.16 standards ('WiMAX') offer improved non-line-of-sight performance and should be able to be deployed inexpensively to provide a robust backhaul capability or to provide direct end-user connection. Furthermore, with the emergence of public standards for this technology, the industry will benefit from industry-sized scale and scope economies and learning effects that will lower costs and barriers to adoption (e.g. end-user uncertainty, fear of stranding). For example, the costs of customer premises equipment should fall to levels much closer to the commodity-level pricing that characterizes Wi-Fi equipment today.

Networks evolved to enable resources and ideas to be shared, and nowhere is the original concept of sharing more present than in MANs. Hot spots are great for the mobile individual, LANs connect a group, but MANs bring people into a community and assist them with social, educational and career challenges. MANs reach the highlands and islands to bring university learning to rural students. MANs bring the Internet to villages in the underdeveloped world that are not reached by traditional infrastructure. In more urban areas, metro networks bring civic groups together – such as government agencies across a city or county – and allow the extension of low-cost Internet access to the community.

However, there are two substantial barriers to this community network sharing. First, the cost of leasing or installing fibre across a county would be impossible to fund. Second, the right-of-way required for digging trenches would be impossible to secure.

WiMAX readily address these two barriers. Wireless radio transmissions go where wires cannot go – or go affordably. Leasing fibre optic lines or installing new cabling to connect corporate offices and rural villages alike is an expensive recurring cost. WiMAX eliminates that cost. High-speed point-to-multipoint links offer high bandwidth at a fraction of the price and bring the benefit of fast, easy installation. Networks are installed in weeks, not months, with very limited (if any) right-of-way installation requirements. With a wireless MAN, communities connect and share as never before possible.

Technology choice for MAN

At the metropolitan level, 802.11's widely available and cheap hardware has made it irresistible for some applications that exceed its design brief. As a result, the 802.16 or WiMAX specification addresses 802.11's limitations for service-provider applications to enable broadband wireless access systems.

WiMAX products were built from the ground up to meet the requirements of both large and small service providers. Over the next 12–24 months the BWA industry can expect WiMAX to become the dominant delivery technology for the MAN, and Wi-Fi™ technologies will revert to being used predominantly in LAN, within the home, office or public hotspot location.

Since the 802.16 MAC explicitly supports point-to-multipoint wireless access that must interface with the telecommunications infrastructure, profiles exist to support Ethernet/IP and ATM (asynchronous-transfer-mode) environments. The system also supports ATM-compatible QoS spaced models – an area that 802.11 is currently trying to improve via ongoing work towards 802.11e specification.

As QoS overhead and throughput inevitably compete for finite bandwidth, 802.16 includes numerous strategies to balance these needs, such as the ability for stations to dynamically request additional bandwidth. The frame structure permits the system to adaptively assign burst profiles to uplinks and downlinks, depending on link conditions, providing a real-time trade-off between channel capacity and transmission robustness. Variable-length protocol and service-data units allow the protocol to assemble multiple units into a single burst, saving the physical-layer transmission overhead. Again, fragmentation permits arbitrary-length transmissions across frame boundaries, but 802.16 includes the ability to manage QoS between competing composite transmissions. A self-correcting bandwidth-request/grant mechanism dispenses with the delays that the traditional acknowledgment sequence causes and also improves the QoS metric.

A service provider who elects to deploy WiMAX Forum Certified systems has a formidable advantage. By deploying 802.16 BWA technologies, a next-generation carrier can develop and deploy a very nimble architecture that leverages many of the core value-added features of wireless systems. By their very nature, 802.16 systems are flexible in that the technology can be redeployed quickly, and essential capital is not stranded in the ground.

Second-generation BWA technologies based upon 802.16/HIPER-MAN bring new tools into the hands of service provider engineering and marketing teams. Public and private network operators worldwide seem to have converged on a set of requirements that must be met by successful equipment manufacturers hoping to win their business.

Key to the operator is a system design focused on minimizing the cost of installation. This can be achieved through high system gain, smart antenna technology, MIMO techniques etc., but the objective is to eliminate or minimize the cost of truck rolls to the operator. Another critical system level requirement is a robust scheduler to address 'carrier-class' deployments where thousands of subscriber stations may be deployed within the catchment area of a single base station. In order to provide the breadth of advanced data services desired by many, if not most, operators, it is important that a fully functioned L2–L7 classifier be included within the MAC on the selected BWA system.

The system should also fully support 802.1p/Q tagging and evolve to support MPLS as required by the operator. Many operators are electing to use wireless to support all their services and therefore are looking for base station and subscriber station hardware that provides support for both structured and unstructured T1/E1 traffic.

Finally, the system should have an associated element management system that enables full FCAPS support including a provisioning methodology that meets the varying business models of public and private network operators.

Ultimately, WiMAX products will be successful if the equipment vendor community manages to deliver products that meet the operator requirements detailed above. The key lies in delivering the performance promised at reasonable cost in a timely fashion. Service providers will focus on three key deliverables: the range delivered; CPE cost and form; and the overall ease of use of the system.

On the radio side, the original 802.16 specifications describe operation within the 10–66 GHz band. Because frequencies greater than about 11 GHz demand a line-of-sight path, this option suits point-to-point links to about 50 km. Line-of-sight operation virtually obviates multipath effects to allow channels as wide as 28 MHz that furnish a maximum 268 Mbps capacity.

FDD (frequency-division-duplex) and TDD (time-division-duplex) physical-layer options are available that transmit using QAM techniques. The FDD option permits both half- and full-duplex terminals, but a typical line-of-sight radio system requires that there are no obstructions, such as trees or buildings, within a roughly elliptical window around the direct transmission path.

Obstructions that lie within about 60% of the envelope of this 'Fresnel zone' can severely degrade signal strength. Clearly, this consideration impacts deployment potential, especially for arbitrary locations within metropolitan areas. To ease this situation, 802.16a supports the conventional propagation model within a 2–11 GHz band. Here, OFDM overcomes variable reception delays, intersymbol interference and multipath reflections to ease reception within reflective environments to allow a typical cell radius of about 8 km. Several physical-layer profiles are available, but the WiMAX Forum is focusing on the 256-point FFT (fast Fourier transform) OFDM mode as its prime interoperability target.

For the future, modifications to 802.16 in the shape of 802.16e may extend the technology's reach to passenger transport systems using the 2–6 GHz licensed bands.

Municipal and community networks

Some municipalities have been providing broadband services based on wired technology (cable or DSL modems). Since such investment is associated with large sunk and fixed costs, it is not surprising that few communities have elected to invest in providing municipal infrastructure, and that the communities that have been most likely to offer telecommunication services are those that already have history of providing municipal utility services like electricity.

In many cases, these municipalities have already invested in an advanced data communication network to support its utility service operations. Also, they already have a relationship with residents and businesses in the community and have to maintain customer support and outside facilities maintenance services for its local utility operations, for example electric power distribution plant.

Typically, the municipal electric utility (MEU) owns or has access to outside structures (poles or conduit) which can be used to deploy the last-mile communications infrastructure. Finally, a number of power companies have been experimenting with using the existing power lines to transmit communication signals, called 'broadband over power lines' (BPL).

WiMAX expands the set of technical options, and depending on the local circumstances, may offer a substantially lower deployment cost. MEUs may elect to deploy communication services via wireless technology, instead of wireline. The new wireless technology is also being deployed to support MAN-sized networks in non-MEU communities and by new types of players. WISPs have emerged in a number of rural communities in the USA and abroad. While most of the WISPs are providing retail services directly to end-users, there are some that have adopted an open access/wholesale model.

Finally, a number of municipalities are facilitating entry by investor-owned wireless service providers by providing access to government buildings and schools for antenna siting, and in some cases, by allowing antennas to be placed on street lights and other municipally owned property for wireless technologies based on small cell sites (short-range wireless technologies like Wi-Fi).

For communities that do decide to deploy MAN-sized access infrastructure, the wireless technologies are important because they expand the range of players and technical options for leveraging existing investments, thereby helping to lower the cost of a municipal deployment. For example, wireless can be used to economically extend public access to municipal fibre or to local government intranet backbone services.

9

WiMAX Opportunity

Historically, communications systems have been rigid, forming linear chains of content, channel and device. Conversely, today's communications systems are inherently flexible, allowing for a mix and match relationship between components. This flexibility provides an opportunity for the convergence of established communications systems and the creation of new services (Figure 9.1).

The international telecom equipment market is valued at more than $300 billion, growing at approximately 15 % a year, faster than most other industries. At year-end 2004, revenues from telecom services worldwide were $1.25 trillion, with 1.2 billion GSM subscribers worldwide, representing more than half of all telephone subscribers. Further, there will be more than 3 billion mobile subscribers in the world by 2010, more than double the current subscriber levels (Figure 9.2).

Broadband is becoming a necessity for many residential and business subscribers worldwide. There are close to 150 million broadband subscribers today, while there were close to 130 million broadband subscribers worldwide at the end of 2004, a 30 % growth from 2003 (Figure 9.3).

High-speed broadband access will be a principal driver of telecom equipment revenue over the next 4 years, helped by increased government support and a stronger economic environment. The broadband access revenue will triple between 2004 and 2008, from $33 billion to $101 billion (Figure 9.4).

The fundamentals for continued growth remain sound.

The Business of WiMAX   Deepak Pareek
© 2006 John Wiley & Sons, Ltd

Million US$

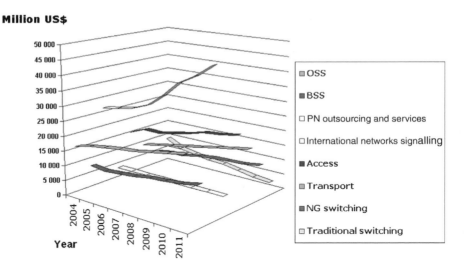

Figure 9.1   Global telecom spending

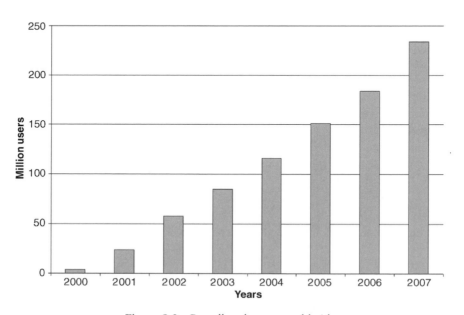

Figure 9.2   Broadband users worldwide

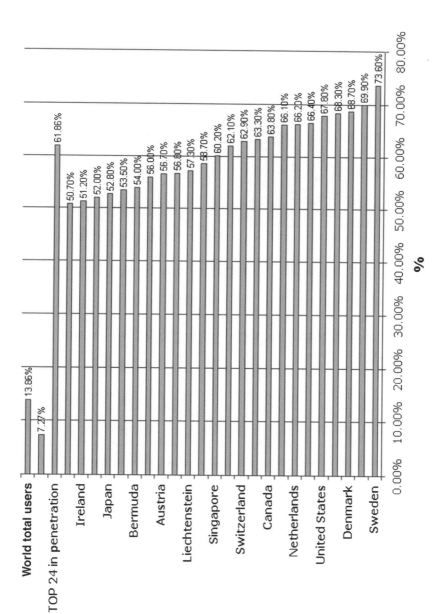

Figure 9.3 Broadband user penetration

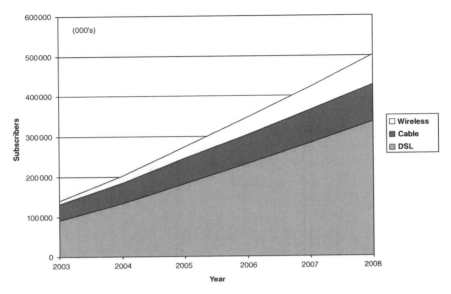

Figure 9.4   Worldwide broadband subscribers according to technology used

9.1   BROADBAND WIRELESS TRENDS: BUILDING MOMENTUM

The momentum of broadband wireless is starting to build. Technologies such as local multipoint distribution system (LMDS) and MMDS emerged in the market with much fanfare a few years ago – but without much success. However, vendors continued to develop second-generation products to eradicate major obstacles, including line-of-sight problems and expensive, highly technical installation. New products and applications have reversed this trend (Figure 9.5).

Thanks to technological advances and the emergence of WiMAX, broadband wireless is gaining momentum. Technological advancements are a key reason for the take-off of BWA infrastructure in recent years. The emerging WiMAX standard will probably play a role in making wireless broadband access more widespread than anything else before it.

Regardless of WiMAX, the BWA industry is growing rapidly. Wireless broadband growth driven by increased broadband take-up and fixed/mobile convergence (mobile telephony seen as a substitute for fixed telephony) is expected to expand globally at a very healthy 27%

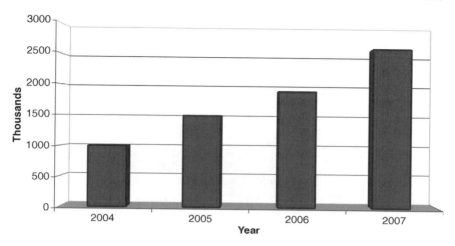

Figure 9.5 Fixed BWA users worldwide

compound annual growth rate (CAGR) between now and 2008 with 10 million BWA customers worldwide by then.

Revenues from services delivered via BWA had already reached $323 million in 2004 and are expected to jump to $1.75 billion by the end of 2006. The market for wireless services, including Wi-Fi and WiMAX, is expected to reach $12.4 billion by 2010. This is significantly greater than the prevailing single-digit growth rate of the telecoms industry as a whole, and this does not take into account the effect WiMAX may have on expanding the market. Wireless broadband can complement existing services in rural and metro areas and is ideal for emerging markets (with less infrastructure etc.). The hottest regions for BWA growth in the years ahead are Asia and Central and Eastern Europe.

BWA infrastructure revenues are expected to reach $5 billion in 2005; revenue from spending on BWA capital expenditures will reach an estimated $22.3 billion in 2005, climbing to $29.3 billion by 2008, a 7.1 % compound annual gain.

Spending on services in support of the BWA infrastructure (including Wi-Fi and WiMAX) such as basic services and support (e.g. field maintenance and repair), professional services, and depot repair and logistics rose by 13.6 % in 2004, rebounding from the 31.8 % drop in 2003 associated with the drop in wireless infrastructure spending (Figure 9.6).

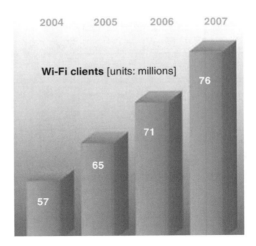

Figure 9.6 Wi-Fi client units worldwide

Wi-Fi Trends

Wi-Fi, which includes wireless standards 802.11, 802.11a, 802.11b, and 802.11g, represents a small but quickly growing component of wireless communications services. Spending on Wi-Fi equipment rose by 31.8 % in 2004 to $4.35 billion. The proliferation of Wi-Fi access points (hot spots), expansion of wireless corporate LANs and growth in the use of network interface cards (NICs) as standard equipment in laptops are the reasons for this boost in spending. Maintaining the trend, spending on Wi-Fi infrastructure equipment is expected to total $7 billion in 2008, a 12.6 % compound annual increase.

Spending on Wi-Fi services is expected to reach $45 million in 2005, higher than $21 million in 2004, and will continue to climb at a 100 % compound annual rate to $335 million by 2008.

Although experiencing healthy growth on a percentage basis, aggregate revenue is expected to remain relatively low because most Wi-Fi services are offered either free as a promotion or bundled with other services. Consequently, Wi-Fi is not expected to become a significant source of service revenue by itself. Rather, it is expected to stimulate other revenue by attracting business and by growing the equipment market (Figure 9.7).

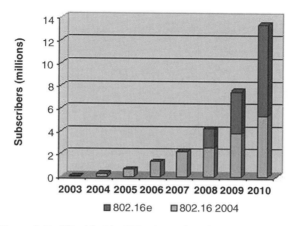

Figure 9.7 Worldwide WiMAX subscribers by standards

WiMAX Trends

Broadband wireless has evolved from an obscure acronym to the next big thing thanks to proponent (Intel, Fujitsu etc.) marketing machine and the formidable progress made by the WiMAX forum, increasing membership to the extent that WiMAX is now synonymous with broadband wireless (Figure 9.8).

The whole concept of standardization is to reduce equipment and component costs through integration and economies of scale that in turn

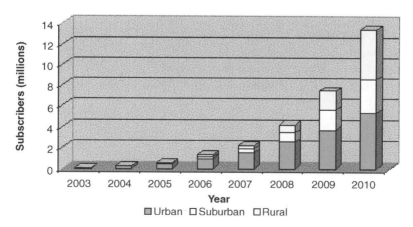

Figure 9.8 Worldwide WiMAX subscribers by demography

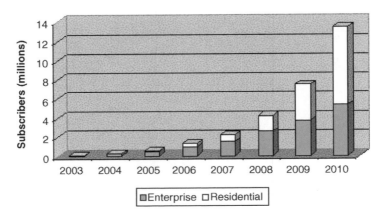

Figure 9.9 Worldwide WiMAX subscribers by segment

allow for mass production at lower cost. In particular, current chipsets
are custom-built for each BWA vendor, making equipment development
and manufacturing both costly and time-consuming.

With large volumes, chipsets could sell for as little as $20 and other
WiMAX components could benefit from these mass volumes as well. We
expect the cost reduction impact to be mostly on the CPE, with an
average selling price of less than $100, and WiMAX will capture 36 %
of the total broadband wireless market by 2008 (Figure 9.9).

Base station costs are more complex owing to the variety of types and
scale. However, base stations are less of a factor in the economic
equation for operator deployments. Industry analysts believe that the
WiMAX market will be worth anywhere from $3 billion upwards by
2009, and a $3.4 billion annual opportunity for fixed and portable
wireless broadband equipment is foreseen by 2010. The 3.5 GHz band
will account for 50% of that multibillion dollar market by the end of the
decade (Figure 9.10).

Fixed Market Trends

The fixed/portable broadband wireless equipment market (sub-11 GHz)
has grown from a $430 million market to a $562 million one – a 30%
increase – and is predicted to pass the $ 2 billion mark by 2010. By 2009
there will be 7.2 million subscribers of fixed WiMAX (Figure 9.11).

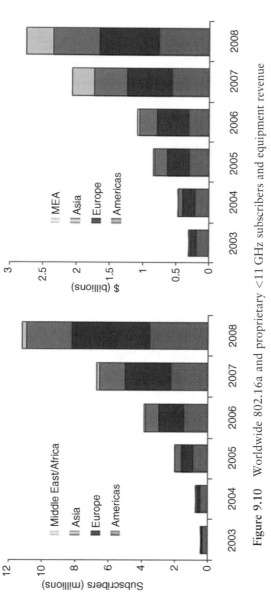

Figure 9.10 Worldwide 802.16a and proprietary <11 GHz subscribers and equipment revenue

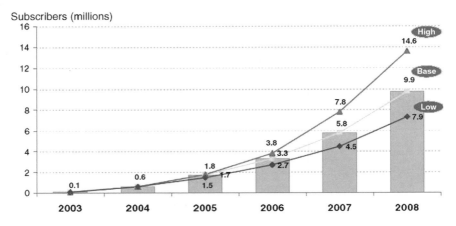

Figure 9.11   Worldwide mobile WiMAX subscribers

Mobility Market Trends

The largest markets for wireless broadband will be for mobile applications. Mobile broadband is being developed from two opposing directions: from the WiMAX side, systems will become increasingly mobile as unification takes place under the 802.16 standard. From the cellular mobile side, systems are being driven to deliver voice, rich media and broadband data over an IP network. Both streams of development eventually will deliver similar data rates. However, currently cellular phone/data network sales greatly exceed BWA in terms of both unit numbers and revenues.

The trend for WiMAX systems starts with the first stage of fixed-nomadic CPEs with systems expected to become WiMAX Certified starting in mid-2005. The second stage of WiMAX systems based on 802.16e will provide greater nomadic ability followed by PCMCIA-enabled laptop mobility (Figure 9.12).

Industry Players

The BWA industry structure is changing rapidly. Several established companies are leaving and new companies are entering the industry. Telecom equipment giants like Nortel, Agere, Marconi and Lucent have all exited the fixed wireless business.

New companies like Alvarion, Airspan and Proxim have entered the industry. They are significant players in developing the new WiMAX

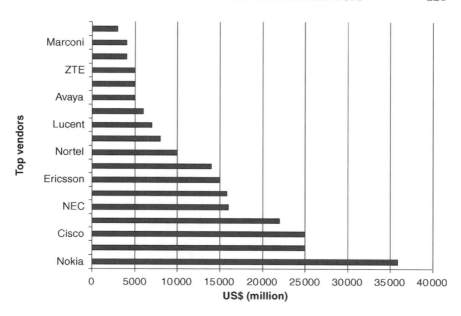

Figure 9.12 Top telecommunication vendors' revenue 2004

standards. As the market grows, the number of vendors is also expected to increase.

Today there are numerous small BWA vendors. To be able to survive they will have to find profitable niches or be acquired. Otherwise, they will eventually die. The market is entering a period of transition and it is expected that the industry will mature from its current niche position and move towards consolidation. WiMAX has split the vendor community into two not-so-evenly divided camps. Because BWA vendors see such dramatic (and painful) changes coming, many are lining up against WiMAX: one can expect the chorus to grow louder and nastier in the months ahead. WiMAX adds value for every stakeholder in the BWA industry (Figure 9.13).

Figure 9.13 BWA value chain

- *Component makers* – WiMAX creates a volume opportunity for silicon suppliers.
- *Equipment makers* – there is longer any need to develop every piece of the end-to-end solution. Businesses can innovate more rapidly.
- *Operators* – standards-based equipment means more cost-effective CPEs. Multiple equipment suppliers and lower costs = lower investment risk, quick provision of E1+ level and 'on demand' high-margin broadband. WiMAX is scalable, enables remote management and supports differentiated service levels.
- *Consumers* – there is more choice of broadband access and support for voice, data and video.

Service Providers

WiMAX is anyone's game. WiMAX blurs the line between fixed and mobile service; as such, it has attracted the attention of both fixed and mobile providers.

Fixed providers tend to embrace wireless broadband as a means of extending their network into rural areas as well as stemming the flow from their enterprise T1 clients. Mobile providers look at wireless broadband as a way to gain enterprise clients and increase the revenue per user. Mobile providers also have the infrastructure upon which to overlay WiMAX networks.

Finally, there are the WISPs. These companies tend to be small and operate in the unlicensed bands. We expect a wave of consolidation to begin as the larger WISPs seek to expand their networks and major carriers seek new revenue opportunities.

Technology Providers

WiMAX 802.16 and the later 802.20 represent a substantial effort by major IT and telecom players to expand the commercialization of WBA technologies and to create convergent technologies with seamless handling of voice, data and video. The WiMAX objectives are to standardize the WBA technologies, drive down end-user costs and make WBA solutions available to a broader marketplace. The rapid growth and development of the Wi-Fi technology has caused a 'kick start' for the new standards. Intel and other chipmakers focus their efforts on

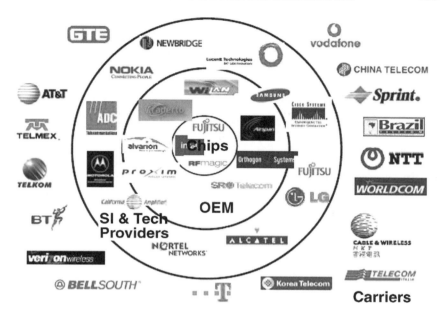

Figure 9.14   WiMAX industry structure

bringing scale to the market. WBA vendors are always interested in less expensive chipsets.

The WiMAX Forum plans to enforce standards compliance among vendor members in the same way that the Wi-Fi Alliance has worked. Compliance to standards results in interoperability, which in turn leads to plug-and-play products. In the coming years WiMAX vendors no longer have to provide end-to-end solutions.

They can specialize on base stations or wireless modems. Specialization will result in competitive pricing and value-added innovations. Nearly all laptops are already Wi-Fi-enabled. An increasing number of PDAs and home entertainment devices, as well as a growing number of cell phones are enabled with Wi-Fi chips. It is expected that WiMAX chipsets will be embedded in laptops and other mobile devices within 2006 (Figure 9.14).

Global Spread

The WiMAX standard will represent a great opportunity for emerging markets, rural areas in developed countries as well as populated areas

Table 9.1 Vendors active in WiMAX space

RF Chip vendors	Service providers	Equipment vendors and OEMS
Atheros	AT&T	Alcatel
RF Magic	BT	Alvarion
RF Integration	China Motion	Airspan
Sierra Monolithics	Covad	Aperto
Skyworks	FT	Motorola
Baseband Vendors	Reliance Infocomm	Navini
Beceem	PCCW	NextNet
Intel	Qwest	Proxim
Fujitsu	Telenor	Redline
Sequans		Siemens Mobile

with a demand for 'hot-spot' wireless broadband services. For the emerging markets this new technology will decrease infrastructure costs significantly. Thus, the Asian and Eastern European markets are expected to experience a rapid growth in the use of WBA technology in the years to come. The same scenario is expected for South America (Table 9.1).

In these markets operators are interested in using WiMAX for low-cost voice transport and delivery. The development of these previously underserved markets will establish economies of scale for the equipment and chipset vendors. Costs will be driven down. This will lead to a fast progression to WBA low-cost connectivity all over the world. WiMAX is expected to bring broadband to the masses. Analysts expect WiMAX to succeed as a new global standard in every geographic market, but for different reasons.

PART Four

WiMAX Strategy

10

Strategy for Success: Service Providers

The extinction of the dinosaurs shows that time brings the unexpected. Natural history is filled with examples of species that did not evolve as fast as their environment changed. What if the dinosaurs could have developed the aquatic capability of a seal, or the hibernation capability of a polar bear or the intelligence of humans? These abilities may have saved them from extinction. Species extinction shows that, for all its power, evolution is myopic and lacks the flexibility to change quickly.

Modern organizations evolve like species in nature, but unlike natural species, organizations can innovate and respond to external shocks – faster and with greater foresight. Now more than ever, innovations in communications technology are changing the environment in which organizations compete. These innovations not only present organizations with new challenges but also provide them with new opportunities.

10.1 UNDERSTANDING DIVERSITY: SERVICE PROVIDERS

Service providers consist of traditional telecommunications companies – incumbent local exchange carriers (ILECs), competitive local exchange carriers (CLECs), ISPs, cable operators and wireless carriers (cellular

The Business of WiMAX   Deepak Pareek
© 2006 John Wiley & Sons, Ltd

and mobile service providers) – as well as relative newcomers to the market, such as satellite companies, wireless companies, developers of new wireline broadband technologies and fibre deployment companies.

Incumbent Local Exchange Carriers

ILECs are wireline telecommunications carriers that own the legacy of telephone network within a geographic area. They offer local telephone service, local toll, long-distance, international, Internet access and broadband services.

Competitive Local Exchange Carriers

CLECs are wireline carriers that are authorized to compete with ILECs to provide local telephone services. They often package their local service offerings with local toll, long-distance, international, Internet access, cable and/or video services. Generally CLECs are not required to duplicate ILEC local service offerings. They can choose which customers to serve (business, residential or both) and what services to offer. CLECs provide telephone services in one of the three following ways or a combination thereof:

- building network facilities needed to connect themselves to their customers' premises;
- purchasing telecommunications services from another carrier (typically an ILEC) at wholesale rates and reselling those services to their own customers at retail rates; and
- leasing parts of the ILEC network, referred to as 'unbundled network elements' (UNEs).

Some ILECs also operate as CLECs outside their original service territories.

Wireless Carriers

Wireless carriers, also called cellular and mobile service providers, are those involved in provision of cellular or mobile telephone services. These carriers also provide high-speed Internet service using mobile wireless technology.

Internet Service Providers

ISPs are independent network operators or ILEC or CLEC subsets involved in provision of Internet services. ISPs also deliver broadband services, generally by purchasing unbundled local loops and providing their own electronics at each end to provide a DSL service to customers. ISPs traditionally have not provided voice services, although some are now offering VoIP telephony.

Cable Providers

Cable companies provide broadband services over their coaxial cable networks. Cable providers are generally granted exclusive franchises by the jurisdiction in which they operate. Cable broadband providers serve primarily residential customers, since many homes across the nation already subscribe to cable video.

Satellite broadband providers

Satellite providers can deploy a broadband service to customers in almost any part of the world. Customers must install a satellite dish with a clear line-of-sight view of the southern sky. So far, it has been a popular choice for customers in rural and other areas lacking an existing broadband infrastructure, where deployment costs are often too high for other broadband providers to enter the market. Deployment costs are substantial, as they involve placing a new satellite into orbit. Satellite providers often set limits on data downloads, with surcharges applied if a customer goes over his or her quota.

Wireless Internet service providers

Some ISPs are now providing high-speed Internet using wireless solutions and are referred to as wireless ISPs. These operators provide high-speed Internet services using fixed or mobile wireless solutions. Fixed wireless technology can offer services to large geographic areas with a modest investment. It is a particularly attractive form of broadband in rural areas, smaller towns and remote areas.

Broadband overbuilders

Broadband overbuilders are a new type of telecommunications provider. Unlike local telephone and cable television companies, which have adapted their existing networks to provide broadband, these providers focus on a core business strategy of building new fibre-optic networks which they use to provide local telephone, cable television and high-speed Internet services.

10.2   STRATEGY DEVELOPMENT

The traditional tools for strategic analysis are not adequate for the task because the challenges that must be modelled today are not static. The problem is not in assessing the advantages and disadvantages of a given technology today. The problem is in anticipating what those wireless technologies will become in the future.

These urgent challenges demand a powerful and illuminating theory to guide decision-making and action. The challenges faced by telecommunication companies of all stripes indicate the need for a strategic framework which can be successfully deployed in a variety of business contexts, providing profound insights into the nature of innovation, competition and industry transformation.

New entrants have typically been the victors when it comes to exploiting disruptive innovations of all colours. As new entrants start generally with a handicap, they need to be more innovative and are inclined to explore business or technology disruption in order to make space for themselves. So the success strategy here is simple: keep doing what successful new entrants have always done.

That is not to say success will be automatic. For example, many CLECs fell into the trap of targetting large, lucrative enterprise customers – the very same enterprise customers the incumbents had to retain to ensure their own survival. Eventually, the incumbents did retain these enterprise customers (by bullying competition in some cases), while entrants burnt their fingers. In essence, the CLECs picked a bar fight with an opponent in adverse conditions.

The fundamental logic for the creation of new, state-of-the-art strategies for a successful performance that meets the market challenges created by WiMAX are:

- Strategy development – conception of strategic options by assessing market potentials from a financial as well as a technical point of view. A suitable method is value chain analysis.
- Strategy evaluation – test of the strategic conceptions and hypotheses for plausibility and feasibility which can be a check against the market challenges and against the corporate strategy pursued so far. Determination of necessary capabilities and other important factors can be done using SWOT and VRIO analyses, which are tools to assess a company's strengths, weaknesses, opportunities and threats as well as its qualities that are valuable, rare, costly to imitate and actually exploited by the organization.
- Strategic positioning – selection of a strategy and the identification of options for the strategic position. A useful tool in this context is the strategy matrix.
- Strategy recommendation – a conclusive concentration of the above-mentioned exercise, which defines the ground rules for the players.

Prescription to Service Providers

New entrants

The prescription for new entrants is therefore to make absolutely sure that their products and services constitute disruptive innovations. They must establish a firm foothold, either in a new market, or in a low-end market, that will pay them to deliver the kinds of improvements that can eventually lead to disruption of mainstream markets. By exploiting the asymmetries of motivation that define competition between disruptors and incumbents, new entrants can enjoy a high probability of success, even in a capital-intensive, scale-sensitive industry such as telecommunications.

Incumbents

Incumbents face a more difficult challenge. Their future depends on their ability to continually deliver a stream of sustaining innovations that their most important customers value. Staying on the industry's sustaining trajectory of innovation is critical not merely to success but also to survival. However, simply surviving is not what shareholders reward. Profitable growth has to be the objective of every senior management team. Since every sustaining trajectory must eventually run out of room to grow, persistent profitable growth depends on finding and exploiting disruptive innovations. So the eternal struggle between present profitability and future growth can be re-cast as the tension between sustaining and disruptive innovations.

WiMAX: Threat or Opportunity

In the race to deliver broadband to consumers and businesses, many wonder if WiMAX will be another arrow in an incumbent's quiver or a new weapon employed by nimble competitors. In the dawn of this new market, traditional wireline carriers and wireless operators are trying to determine how – if at all – to embrace this technology.

Key to this development will be careful planning by the incumbents to determine how to best utilize this emerging wireless technology – just as they begin to analyse moves with Wi-Fi, another disruptive technology.

Technologies make many things possible, but if they are not in demand, they may never become adopted or even noticed. Technologies can also create production efficiencies and cost savings, but earning a fair return on large upfront investments requires that firms change the way work is organized around technologies.

In today's rapidly developing, technologically connected world, organizations have the power to shape the future and adapt to their changing environments. In order to survive, organizations must build new relationships with outside players, be in a position to observe the revelation of information and be armed with a rich set of choices from which to select the appropriate response.

However, in many instances the future choices themselves are far from known. In other words, much of the value comes from an ability to gain proprietary learning from strategic experimentation. Technology presents today's organizations with valuable opportunities and difficult challenges in the face of tremendous uncertainty. Therefore today's organizations must evolve flexibly and intelligently or face extinction.

The WiMAX revolution is only beginning. As with any new standard, the first inclination of some will be to deploy the new technologies right away – risking instability and additional overhead – to be first to market. Others are likely to move with caution, waiting for a stable WiMAX solution at the risk of losing market share.

The most successful carriers will be those who navigate this period of risk and reward to generate efficient growth with these new wireless services. Finding the right WiMAX-ready platform for deployment today can provide a balance of performance and reliability to enable success. WiMAX is a great base upon which to build equipment, so carriers are faced with a strategic decision today: to deploy WiMAX or not to deploy WiMAX.

WiMAX Adoption

WiMAX is of interest for incumbent, alternative and mobile operators. The incumbent operators can use the wireless technology as a complement to DSL, allowing them to offer DSL-like services in remote, low-density areas that cannot be served by DSL. For alternative operators, wireless technology is the solution for a competitive high-speed Internet and voice offering, bypassing landline facilities, with applicability in urban or suburban areas. The larger opportunity will come with portable Internet usage, complementing fixed and mobile solutions in urban and suburban areas. Therefore it will enhance the business case for mobile operators by giving a high level of bandwidth access to a large number of potential end-users.

In 2004–2005 a number of operators deployed wireless broadband networks, and many are planning to move to the much-hyped WiMAX standard when hardware becomes available in 2006. Early network deployments have highlighted a number of business models. Some networks have been deployed by start-up operators looking to compete directly with existing fixed infrastructure providers, while others have focused on servicing a specific mobile broadband segment. WiMAX technology is expected to be adopted by different incumbent operator types, for example, WISPs, CDMA and WCDMA and wireline broadband providers. Each of these operators will approach the market based on their current markets and perceived opportunities for broadband wireless as well as different requirements for integration with existing (legacy) networks.

As a result, 802.16 network deployments face the challenging task of having to adapt to different network architectures while still supporting standardized components and interfaces for multivendor interoperability. For the first time, broadband wireless operators will be able to deploy standardized equipment with the right balance of cost and performance, choosing the appropriate set of features for their particular business model.

10.3 WIRELINE CARRIERS

The dramatic increase in the demand for data-intensive solutions for Internet access, e-commerce, tele-working and multimedia applications, together with the telecommunications deregulation, is driving the requirement for increased broadband capacity. These escalating band-width demands are impelling service providers to broaden Internet access provision for both business and home users alike.

To date, carriers have relied on wire-based solutions in order to provide their customers with broadband access. Those solutions have generally been based on upgrading their existing copper twisted-pair infrastructure utilizing ISDN and DSL technologies. However, DSL can reach only a small portion of the market as it is limited by the length and quality of the copper wire. DSL requires high-quality copper and short distances from the switchboard as well as highly skilled installers.

In addition, these solutions are not cost-effective and are slow to deploy. These difficulties have led to limited roll-out and uptake of DSL. In many countries, the slow unbundling and roll-out of DSL have led alternative providers, frustrated at the difficulties associated with gaining the necessary access to the incumbent provider's local loop, to seek an alternative means of providing their customers with voice, data and Internet access services. Carriers are seeking a cost-effective technology that will address the need for the fast deployment of broadband access.

BWA is gaining increased broadband market share as it overcomes the restrictions of wire-based technologies, in terms of deployment costs and time. Overcoming the restrictions of 'old copper', lack of infrastructure and asymmetry (i.e. higher data rates downstream than upstream), BWA provides a cost-effective alternative that is quick to deploy, symmetrical and flexible to build-out. BWA solutions provide cost benefits in roll-out, including lower capital cost and the fact that capital expenditure is incurred incrementally. In addition, BWA enables accelerated time to market, for quick provisioning to new customers, permitting demand-based build-out.

Wireline carriers and cable/DSL operators see WiMAX as a way to break into the wireless market without taking the (expensive) cellular route, or plan to use 802.16 as an extension to their existing services in areas that are hard to wire. For those looking to fill the holes in their cable or DSL networks, wireless is the most viable solution.

Wireline carriers and service providers generally, including incumbents and CLECs, could find WiMAX very attractive for parts of their

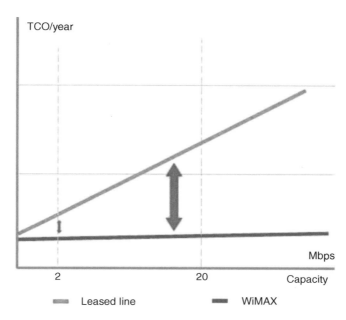

Figure 10.1 WiMAX as backhaul

networks and for specific applications, such as backhaul from remote sites, for example cellular base stations (Figure 10.1).

Service providers seeking to build broadband infrastructure in developing countries should take fibre as far as they can, then go wireless. WiMAX will allow these companies to deploy more quickly and at lower costs. WiMAX has the potential to be the great equalizer in broadband access. In short, bridging the gap of the digital divide has never been more within our reach.

There is also a wider picture, in which WiMAX becomes an enabler of a more seamless mode of communications, essentially built on Ethernet. In this, WiMAX provides blanket Ethernet coverage of an urban, suburban or rural area to make broadband ubiquitous.

New services are critical for service provider success moving forward. As voice and legacy data service revenue declines, service providers need new services to fill the revenue void. More importantly, enterprise customers are demanding new services from their service providers. If the incumbent provider does not have what customers need, they will redirect their money to a new service provider who does offer it.

Enterprises are demanding the following functionality and requirements from their service providers:

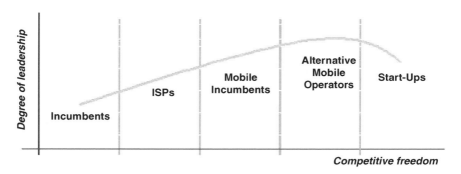

Figure 10.2   Leadership and competitive freedom matrix

- higher-availability data services;
- preservation of their existing data and voice services;
- interworking between their existing data services and new data services;
- lower price per Mbps of data transported;
- LAN-like functionality in the WAN;
- strong and specific SLAs for premium services (for which they are willing to pay more);
- tiered options in terms of both bandwidth and SLAs so that services can be custom-tailored to their specific (and changing) network requirements;
- large footprint so that all major and remote sites can be linked together without requiring a hodgepodge of different carriers and services.

Metro Ethernet is one new data service that is gaining momentum. Storage extension is another area of growing interest within the enterprise, as enterprises need to connect datacentres across distances ranging from tens of miles to thousands of miles, primarily for business continuity and disaster recovery purposes.

Fixed-line players who have been kept out of the wireless market because of regulatory fiat are likely to be excited at the prospect of getting a piece of the wireless cake (Figure 10.2).

10.4   NEW ENTRANTS

BWA solutions are becoming increasingly popular as operators discover their ease of deployment and build-out. A wireless network can be deployed in parallel with the existing infrastructure or as a complete infrastructure for emerging operators.

Research has shown that the cost of deploying the last mile will continue to fall for wireless networks while it remains constant for copper-wire networks. Furthermore, WiMAX needs minimal access equipment to support a large number of subscribers – a single base station can support thousands of subscribers.

WiMAX provides the optimal choice for new operators seeking cost savings and rapid deployment, enabling them to competitively win customers through attractively priced services. WiMAX is especially well suited to new entrants as it provides the best opportunity to enter the data and VoIP market without dependence on incumbent providers:

- it is possible to control the infrastructure;
- WiMAX is cost-effective and quick to deploy;
- capacity can be increased as demand grows.

New operators and wireless service providers see WiMAX as a way to break the cost barrier in the broadband market. However, wireless broadband may be more important to new rivals of the big mobile operators. Those new players could include cable operators who want to compete with telecommunications carriers by offering a suite of TV, Internet access, wireline phone and mobile voice and data.

10.5   ALTERNATIVE CARRIERS

The improved deployment costs equation will spur alternative telecommunications carriers to turn to WiMAX technology as a viable business model. Further cost reductions will accelerate WiMAX deployment in later stages, creating a legitimate option beyond traditional copper wires and wireless (mobile or cellular) for voice service. As evidenced by the mass exit of long-distance carriers from consumer markets, rising access costs have reduced opportunities within local markets for nonincumbent operators.

Leasing local copper lines from ILECs has proved to be a flawed and unworkable business model for long-distance carriers. With the departure of long-distance operators from the consumer market, the local voice competition became intermodal, emphasizing the battle between wireless and wireline carriers, and to some extent between DSL and cable broadband operators. WiMAX will enable alternative carriers to regain a foothold in the market and offer consumers another service option.

The economics of the 'wireless local loop' will drive alternative carriers to leverage WiMAX technology, resulting in enhanced competition in the consumer voice market. Wireless economics have already been proved correct in mobile voice services, where the cost per line is 40 % of the cost of equivalent wireline services. These same economies will hold true in wireless data using WiMAX.

Although WiMAX does not create a new market, it enables standardization of the technology required for the volume economics that reduce costs and enable broader market growth. WiMAX will prove economically beneficial for alternative carriers in the following key areas:

- Reduction of capital expenses – by 2008, the total capital cost per customer will be less than $100.
- Reduction of operational expenses – operational costs will be cut by nearly 41 % compared with current wireline operation costs.
- Reduction of customer turnover – by emphasizing centralized deployment and customer self-service, carriers will reduce turnover through increased customer satisfaction and will reduce expensive truck rolls by 53 % from current levels.
- Service differentiation – current fixed broadband offerings cannot provide mobility. With the explosion of VoIP during the same period, mobility will become increasingly important. With WiMAX and VoIP, operators will be able to offer a voice service for both fixed lines and mobile users in a metropolitan area.

WiMAX could well be the technology that breaks the local-loop monopoly for adventurous and well-funded ISPs wanting to differentiate themselves and escape from their dependence on the local teleco. Such carriers require:

- increased reliability of services (less than 99.99 % availability from ILEC);
- shorter deployment time frames;
- capacity upgrades;
- platform for migration to IP services;
- cost-effective alternative to ILEC;
- market-wide solutions;

- carriers in search of:
  - 'lower risk' alternatives to ILEC;
  - converged voice and data services solution;
  - ubiquitous coverage;
  - competitive prices.

10.6 CELLULAR AND MOBILE SERVICE PROVIDERS

The ongoing goal of cellular services providers has been to make networks faster to enable new revenue-producing Internet access and multimedia and data-based broadband services in addition to telephony. For example, carriers want to offer mobile Internet services as fast as those provided by cable- and DSL-based wire-line broadband technologies. This process has taken the industry through various generations of radio-based wireless service: after the first generation (1G) analogue cellular service, they have offered 2G, 2.5G and, since 2001, in some fortunate (or maybe unfortunate) part of the globe, 3G digital technology. However, 3G has disappointed many in the industry because of its high implementation costs and slow adoption, and because its initial deployments did not support services that carriers wanted to offer.

As carriers upgrade their 3G offerings, they are looking perhaps 5 years ahead to 4G services, which would be based on the Internet protocol and support mobile transmission rates of 100 Mbps and fixed rates of 1 Gbps. Presently the subject of extensive research, 4G would enable such currently unavailable services as mobile high-definition TV and gaming as well as teleconferencing.

Wireless companies are thus preparing the transition to 4G from 3G, which could include 3.5G technologies, as the 'cellular technology hits the road again'. Some carriers are looking at new technologies such as IEEE 802.16 and IEEE 802.20. Many providers, though, are simply upgrading the wireless technology they are already using to avoid changing their networking infrastructure. However, this would continue the current problematic situation in which providers throughout the world work with incompatible cellular technologies (Table 10.1).

Four Cellular Paths

There are four categories of next-generation wireless technologies, typically implemented via chipsets, radio transceivers and antennas.

Table 10.1 Comparison of cellular technologies and WiMAX

	Cellular			WiMAX	
Metric	Edge	HSPDA	1 × EVDO	802.16-2004	802.16e
Technology family and modulation	TDMA GMSK and 8-PSK	WCDMA (5 MHz) QPSK and 16 QAM	CDMA 2K QPSK and 16 QAM	OFDM/OFDMA QPSK, 16 QAM and 64 QAM	Scalable OFDMA QPSK, 16 QAM and 64 QAM
Peak data rate	473 kbps	10.8 Mbps	2.4 Mbps	75 Mbps (20 MHz channel) 18 Mbps (5 MHz channel)	75 Mbps (max)
Average user throughput	T-put <130 kbps	<750 kbps initially	<140 kbps	1–3 Mbps	80 % performance of fixed usage model
Range outdoor (average cell)	2–10 km	2–10 km	2–10 km	2–10 km	2–7 km
Channel BW	200 kHz	5 MHz	1.25 MHz	Scalable 1.5–20 MHz	Scalable 1.5–20 MHz

GSM

The 2G global system for mobile communications (GSM) technology, implemented in much of Europe and Asia, is based on time-division multiplexing. In GSM, TDM increases bandwidth by dividing each cellular channel into eight time slots, each of which handles a separate transmission. The channel switches quickly from one slot to the other to handle multiple communications simultaneously.

GSM-based wireless services include 2.5G general packet radio service; 2.5G enhanced data GSM environment (EDGE); 3G wideband CDMA (WCDMA) used in the Universal Mobile Telecommunications System (UMTS); and 3.5G High-Speed Downlink Packet Access. All are currently in use or are in trials.

WCDMA, currently deployed in Europe and Japan, uses a 5 MHz-wide channel, which is big enough to enable data rates of up to 2 Mbps downstream. The technology also increases GSM's data rates by using higher capacity CDMA, instead of GSM's usual TDMA, modulation techniques. However, WCDMA uses different protocols from CDMA and is thus incompatible with it.

HSDPA uses a higher modulation rate, advanced coding and other techniques to improve performance. Otherwise, HSDPA is similar to and can be implemented as a software upgrade to a WCDMA base station as Vodafone spokesperson Janine Young said. Both technologies operate in the 2.1 GHz frequency range. HSDPA offers theoretical and actual download rates of 14.4 and 1 Mbps, respectively, but it addresses only downstream transmissions. HSDPA networks use existing UMTS approaches for the network's uplink. Thus, HSDPA supports applications that primarily require one-way high-speed communications such as Internet access but does not support two-way high-speed communications such as videoconferencing. High-speed uplink packet access technology will provide faster uplink speeds when finalized.

CDMA

The 2G code-division multiple-access technology, developed by Qualcomm and used primarily in the USA, does not divide a channel into sub-channels, like GSM. Instead, CDMA carries multiple transmissions simultaneously by filling the entire communications channel with data packets coded for various receiving devices. The packets go only to the devices for which they are coded.

CDMA-based wireless services include the 2G CDMA One and the 3G CDMA2000 family of technologies. All CDMA-based approaches operate at the 800 MHz or 1.9 GHz frequencies.

CDMA2000 1x, sometimes called 1xRTT, is at the core of CDMA2000. It runs over 1.25 MHz-wide channels and supports data rates up to 307 kbps. While officially a 3G technology, many industry observers consider 1xRTT to be 2.5G because it's substantially slower than other 3G technologies.

CDMA2000 1xEV (Evolution) is a higher-speed version of CDMA2000 1x. The technology consists of CDMA2000 1xEV-DO (data only) and 1xEV-DV (data/voice).

EV-DO separates data from voice traffic on a network and handles the former. Initial versions support theoretical maximum rates of 2.4 Mbps downstream and 153 kbps upstream. Recent revisions to EV-DO and EV-DV will support 3.1 Mbps downstream and 1.8 Mbps upstream theoretical maximum rates. Real-world rates are about half as fast.

The latest EV-DO revision reduces the maximum transmission latency from 300 to 50 ms, making it more suitable for Internet telephony, which requires near-real-time responses. However, the new revision will not be ready for implementation until 2008.

The International Telecommunication Union (ITU) recently approved the 3.5G CDMA2000 3x, also called CDMA Multicarrier. CDMA2000 3x will use a pair of 3.75 MHz-wide channels and is expected to provide high data capacity with transmission rates of 2–4 Mbps. No companies are deploying CDMA2000 3x yet.

IEEE 802.20

An IEEE effort led by Flarion Technologies and supported by vendors such as Lucent Technologies and Qualcomm is developing a cellular standard – based on Flarion's Flash-OFDM – that could handle voice, multimedia and data.

IEEE 802.20 will be a packet-switched technology operating between 400 MHz and 3.6 GHz that could offer optimal data rates of 6 Mbps downstream and 1.5 Mbps upstream.

Flash-OFDM, implemented primarily in Europe, works with OFDM and fast-frequency-hopping spread-spectrum technology, which repeatedly switches frequencies during a radio transmission. This sends a signal across a much wider frequency band than necessary, spreading it across more channels on a wider spectrum and increasing signal capacity.

In real-world systems, the average user would experience about 1 Mbps of downstream and 500 kbps of upstream bandwidth. Major vendors are presently focusing on other cellular technologies, which have put IEEE 802.20 on the back burner.

WiMAX

WiMAX (worldwide interoperability for microwave access) technology, based on the IEEE 802.16 standard, promises global networks that could deliver 4G performance before the end of this decade.

IEEE 802.16d, approved last year, has received support from numerous companies including Alcatel, Intel and Samsung. However, the standard supports only fixed, point-to-multipoint and metropolitan-area-network technology that works via base stations and transceivers up to 31 miles away.

WiMAX is fast in part because it uses orthogonal frequency-division multiplexing. OFDM increases capacity by splitting a data-bearing radio signal into multiple sets, modulating each onto a different sub-carrier – spaced orthogonally so that they can be packed closely together without interference – and transmitting them simultaneously.

The IEEE 802.16e Task Group and companies such as Intel expect to finish work next year on 802.16e, which adds mobility to WiMAX by using a narrower channel width, slower speeds and smaller antennas.

The Task Group is still working on various aspects of the standard. Also, WiMAX has yet to implement features such as mobile authentication and handoffs that are pivotal to a mobile network. This will take time and testing. IEEE 812.16e will operate in the 2–6 GHz licensed bands and is expected to offer data rates of up to 30 Mbps. In South Korea, LG Electronics and Samsung have developed a technology called WiBro (wireless broadband). Proponents such as Samsung advocated basing IEEE 802.16e on WiBro. Intel and some other WiMAX supporters opposed this because WiBro uses a different frequency band and carrier structure than IEEE 802.16. An ongoing disagreement could have created two rival technologies, thereby splitting the market and inhibiting adoption. However, the companies on both sides have agreed to merge the two technologies in IEEE 802.16e.

The five reasons why carriers will take an interest in WiMAX are:

- cellular network congestion due to high-speed data;
- multimedia take-up rates;
- spectral efficiencies and cost per bit of transmission;
- operator frequency spectrum strategy;
- the vision of delivering personal broadband.

Despite the considerable investments required to bring WiMAX networks on-line, carriers can make money from WiMAX, without cannibalizing existing business.

If the IEEE completes work on 802.16e this year, we will be able to see it in mobile phones and PDAs in 2007–2008. Wireless carriers that do not have 3G spectrum or a 3G strategy might decide to use mobile WiMAX to leapfrog to 3G.

An intriguing question is where this potentially seamless WiMAX world leaves mobile operators, especially those implementing 3G. Portable WiMAX-enabled devices could easily outperform 3G devices for many bandwidth-intensive applications, especially those oriented towards business, where a common Ethernet environment would be very attractive.

3G Wireless Carriers

3G cellular operators may embrace wireless broadband technologies to supplement their 3G offerings. Wireless broadband might complement 3G in areas where there is high demand for wireless data, such as city centres.

10.7   NEW REVENUE OPPORTUNITIES

WiMAX offers wireless and wireline service providers new revenue opportunities. Wireless carriers can use WiMAX to provide fixed telecommunications services to businesses and residences, using seamless mobility applications to offer unified service and billing. These services include T1/E1 or faster services using IPSec quality of service to maintain service level agreements (SLA) for businesses, while making excess capacity available to residences to compete with DSL and cable broadband ISP. Wireless carriers are very well positioned to offer this

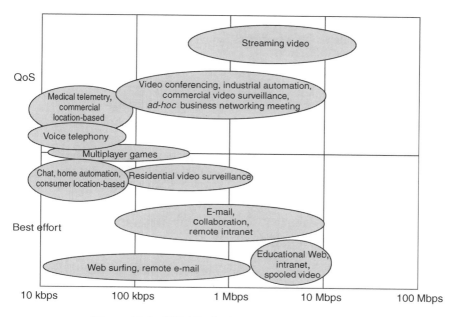

Figure 10.3 WiMAX high QoS: more applications

type of service thanks to their extensive portfolio of cell sites and existing customer marketing relationships.

Wireless carriers can use WiMAX as a less expensive way to backhaul cell sites, especially 3G cell sites, which will rapidly require multiple T1 or E1 links as 3G data traffic increases. This backhaul can be carried in either the licensed or unlicensed spectrum, and since it is fixed at both ends, it can use WiMAX standards in licensed bands.

Wireless and wireline carriers can use WiMAX to backhaul public WiFi (802.11) hot spots, and to extend coverage. The business case for public WiFi has not been highly successful so far, mostly because of the high cost of backhauling the many sites required due to WiFi's poor coverage. The only successful carrier models so far have been using private hot spots (in coffee houses, for example), where the backhaul costs are covered by private operational requirements, and then excess capacity is leased to a carrier. The use of using inexpensive wireless backhaul on WiMAX has the potential to change the WiFi business model.

Existing wireless carriers can use WiMAX revision E as an alternative to cellular 3G deployments, especially where the 3G spectrum has not been allocated, or has not been won by a 2G operator. WiMAX telephone devices that support VoIP are on the drawing board from Motorola's mobile devices business, as well as from other vendors. Intel expects that notebooks will begin to incorporate WiMAX technology during 2006 or 2007, and handsets for mobility will be available by 2007 or 2008. WiMAX's superlative RF performance and low cost of deployment make it a viable and lower-cost alternative to using 3G spectrum for data downloads, or range-limited WiFi hot spots.

Greenfield carriers can use WiMAX revision E to deploy competitive wireless mobility networks using high-frequency spectrum. The superior link budget performance of WiMAX allows operators in higher spectrum bands such as 3.5 GHz to compete well with those in traditional 2G and 3G spectra, and VoIP WiMAX devices will support voice telephony as well as data-centric multimedia applications.

Cable and DSL operators can use WiMAX as an alternative to laying additional copper lines, while making use of their existing IP core networks. These operators can also extend their service areas into hard to reach areas, such as rural areas (taking advantage of government rural broadband subsidies to do so), or into other territories for which they lack wireline presence. For example, in many countries cable companies and telephone operators enjoy a monopoly in one territory, but are locked out of others because competing companies own wireline operating rights in those territories. WiMAX offers a way to offer broadband services in these other territories using wireless.

10.8 VALUE ADDED SERVICES – THE BUSINESS IMPERATIVE

For service providers around the world, the pressure is on. While revenue from traditional voice services still accounts for a significant portion of carrier income, that revenue is steadily shrinking due to increased competition associated with deregulation and the prevalence of wireless substitution. Meanwhile, traditional or 'basic' broadband services are not doing enough to offset these service provider revenue declines. In the case of DSL-based services, competitive flat-rate prices often cannot cover the capital investment and operating, regulatory and third party expenses required to deliver and maintain basic broadband access.

Today incumbent providers are losing on average $2–4 a month per subscriber on DSL in most of the developed world, including North America and Europe. In these markets the ILEC needs 3.5–4 million subscribers to make its DSL service business model work. As of year-end 2004, ILEC in these markets had on average just over 2 million DSL subscribers. With cable multiple systems operators (MSOs) now targeting the voice-data-video 'triple play' in broadband services, incumbent service providers must look beyond basic Internet access, beyond the delivery of 'fat pipes', to stave off the competition, to achieve sustainable broadband profitability and to increase average revenue per user (ARPU) or, more importantly, average margin per user (AMPU).

Generally, access fees account for approximately 90 % of revenue earned from broadband users while services such as premium content represent just 3 %. If service providers want to realize broadband potential, they need to shake these numbers up. They need to move broadband services up the value chain. With next generation network infrastructure largely in place and high-speed access penetration growing, now the broadband challenge is offering value-added services that leverage existing and emerging broadband access technologies, whether xDSL, fibre-to-anything (FTTx) or even Wi-Fi or WiMAX innovations.

Premium Services – Key to Survival

Incumbent operators need to move ahead and provide on-demand and pay-per-use services that attract new customers expand markets and – perhaps more importantly – open new revenue streams for current subscribers, with relatively minimal incremental costs to the service provider. The future of broadband is not merely Internet access, it is:

- premium mass market consumer services such as ultrahigh-speed Internet, online gaming, streaming audio and broadcast high-definition interactive TV;
- business offerings for enterprise end users such as VoD, VoIP, Voice over wireless IP (VoWI) and IP PBX connectivity.

Some of these value-added services are available today while others are well on their way.

However, for most service providers, the missing piece in the value-added broadband service puzzle remains cost-efficient operations to support the rapid introduction, profitable volume deployment and ongoing maintenance of these dynamic value-added service offerings and the QoS requirements that go along with them.

Service Delivery Challenges

As carriers look to a new generation of value-added broadband services, however, they must re-examine the service fulfilment strategies that have served well for basic wired or wireless services. So far these services have been suitable for Web surfing, email and file downloads. For consumers, these flat-rate offerings are a step above dial-up service. For businesses, they are a cheaper alternative to private lines.

The question is, can these strategies stand the test of more dynamic service offerings that leverage much of the same access infrastructure?

The introduction of value-added, on-demand and pay-per-use services such as voice telephony, live audio, tele-networking, HiDefTV, live video, video telephony and online gaming means that exception services will become the norm. One size will no longer fit all subscribers, nor all applications all the time. With the emerging generation of broadband services, quality of service and class of service (QoS/CoS) requirements will be much more stringent and much more varied. Service changes may be frequent (e.g. to accommodate fluctuating bandwidth requirements), and in many cases services will be of limited duration (e.g. the time it takes to view a concert), while time-to-service will be critical.

Meanwhile, requirements for new network features will probably increase as service catalogues expand and technology life cycles contract. Service providers are also hoping that the number of service orders will grow as individual subscribers take advantage of multiple service offerings.

With its array of premium content and interactive applications, the new world of broadband brings with it a host of service provisioning requirements that extend beyond many carriers' existing, static service delivery models.

WiMAX: the Solution

Relying on in-house one-size fits all pre-provisioning solutions to meet the demands of next generation broadband services and the underlying network technology is not economically feasible. To realize the potential of next generation broadband-enabled services, providers need to evolve to a more dynamic service provisioning strategy – one which enables them to take full advantage of both existing feature-rich networks and emerging network technologies; one which can speed the rollout and delivery of value-added services and keep pace with their frequently changing service level requirements; and one which tightly contains operations costs, leveraging existing provisioning strategies.

By leveraging the efficiencies and innovative service delivery characteristics of WiMAX, providers can redefine broadband economics, building a stable profit centre based on a new generation of services.

10.9   RECOMMENDATION

New Operators – WiMAX is a dream technology which they should lap up.

Incumbents – all incumbent operators, must assess existing infrastructure as well as services offered, and they must deploy WiMAX to enhance their portfolio as well as the reach of the service.

11

Strategy for Success: Equipment Vendors

The vendor community has been divided into two not-so-evenly distributed camps. Vendors are either with or against it. WiMAX Ultimately, vendors will find it difficult to resist the commoditization of fixed wireless chipsets. They will either embrace standardization and focus their energies on value-added features and services or focus on proprietary equipment geared towards niche applications and markets.

Many operators are eager to test first-generation WiMAX equipment. However, in comparison to what the proponents of WiMAX expected, the number of operators trying the technology is quite low. Over the years, operators have been burned by BWA equipment that did not live up to expectations. This means that WiMAX Forum and its vendor members, including Intel, need to be very aggressive on pricing and promotion to convince a few bold operators to deploy pilot projects.

Price is key to persuading more operators to accept the solution. Price pressure will be intense in this market, even for first-generation products for two reasons, even though the price advantage of WiMAX is also substantial.

The first reason is that, to conquer the inertia of the first mover, the cost level needs to be more appealing and the value more compelling. The Second reason is of the vendors' own creation. The hype created by them over the past year or so has increased the level of expectation from the technology to unrealistic levels. Vendors will have to compensate for this by more aggressive pricing.

The Business of WiMAX   Deepak Pareek
© 2006 John Wiley & Sons, Ltd

First-generation WiMAX products will be complementary to the existing networks, rather than a replacement. WiMAX vendors will be shipping interoperable base stations and CPE by the last quarter of 2005. Despite the delay, in the certification of products designed for fixed wireless links using the emerging WiMAX standards, equipment and component makers are upbeat and already turning their attention to the next generation: mobile WiMAX.

WiMAX vendors see this technology as an alternative to stationary broadband services such as DSL and cable. Most of the vendors are making steady progress on fixed wireless broadband products but look forward eagerly to a future mobile WiMAX as 'the opportunity' (Figure 11.1).

A key to the success of mobile WiMAX will be more unified specifications. Currently the WiMAX Forum industry group defines many implementation options through what it calls 'profiles'. That may work for fixed WiMAX because the customer equipment for that technology stays at the subscriber's home, but it will not work for a mobile technology in which a device could be taken to many different countries.

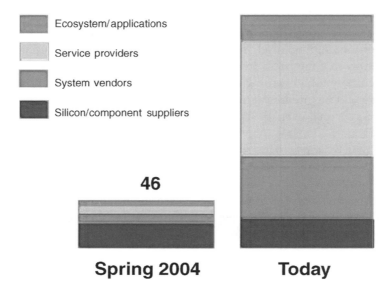

Figure 11.1 WiMAX Forum

Vendors are also pushing for a harmonized allocation of spectrum for WiMAX, with an eye to 2.5 GHz as a band that could be opened to mobile WiMAX in many countries, allowing for international roaming. However, getting many countries to agree on the use of a spectrum band is a tall order. The 2.5 GHz band is assigned to wireless broadband services in much of the Western Hemisphere but not in many other parts of the world.

11.1   WiMAX VALUE CHAIN

WiMAX is very much a global initiative. We now tend to categorize WiMAX applications into four versions or generations – fixed, nomadic, portable and mobile – the former having more complete IEEE standards than the latter.

The key to success in the widespread market adoption of these standards is fostering cooperation between standards bodies, regulators, service providers, manufacturers, equipment vendors, component vendors, system integrators and every other player in the WiMAX ecosystem.

WiMAX Ecosystem

The WiMAX Ecosystem is where the all players in the WiMAX value chain come together to influence the industry's evolution, enabling thousands of future wireless networks worldwide, by bringing together leaders in wireless technology to create a common platform.

The WiMAX Forum plays critical part in supporting this WiMAX Ecosystem and achieving its goals. The WiMAX Forum's principal members tend to be global equipment and component vendors and top-tier operators. Some of the more active members include Airspan, Alcatel, Alvarion, AT&T, BT, Cisco, Clearwire, Intel, Lucent, Motorola, Nextel, Samsung, Siemens, Sprint and ZTE. WiMAX Forum members are dedicated to the delivery of common network architectures and protocols that facilitate consistent deployments of fixed, nomadic, portable or mobile wireless.

WiMAX is an unprecedented globally standardized platform, which even the mighty 3G mobile community was unable to create. For example, a [3G] multimode phone will not necessarily work in Japan or Korea, but a WiMAX device will work anywhere in the world provided the place has a WiMAX network.

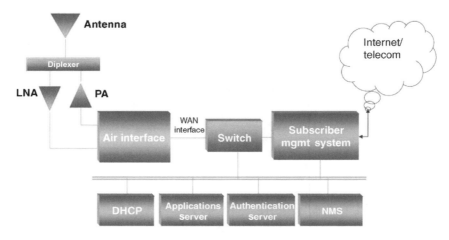

Figure 11.2   Base station component

WiMAX Vendors

WiMAX vendors can be classified broadly as those providing equipment, original equipment manufacturers (OEM), and those providing components (silicon, RF, software, etc.) for this equipment, component manufacturers (CPs).

Among the component manufacturers, the most important players are those providing the chips as, with advent of system-on-chip (SoC) technology, a chip vendor provides the complete hardware and software embedded on the silicon chip.

Further OEMs and CPs can be classified depending on the WiMAX technology segment they serve, the infrastructure, namely base stations, and the CPE (Figure 11.2).

Infrastructure

The infrastructure is are vendors providing equipment or component for base stations. As operators will judiciously launch service in areas in which payback will be quick, the growth in this segment will be slow and controlled. Base station function and its ideal specification differ from those of the CPEs, as does the nature of work going into chip development. BS also cost substantially more than CPE.

Extensive chip-level work is going on in the base-station area. Base-station designs must handle higher aggregate data rates because one base

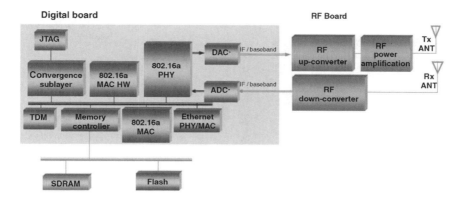

Figure 11.3 Customer premise equipment component

station supports many clients. The higher data rates often mean a different integration strategy; for instance, the combination of MAC and PHY layers may no longer be the best path. Moreover, base stations need to have flexibility, especially in the case of a new standard such as WiMAX; flexibility is often a vital requisite.

Much of the base-station-chip development to date is software or IP (intellectual property). Base-station designs will rely on programmable technologies, including DSPs and FPGAs. For the manufacturers the total cost of chipsets will initially vary around a base price point of $10 000. The big DSP vendors will play an important role in the WiMAX base-station space (Figure 11.3).

Customer premise equipment

A subscriber station or CPE is composed of three main elements: the PHY, which includes a base band, the MAC and an analogue RF front-end that serves as the means of placing signals into a specific frequency band. Equipment vendors look to chip makers to provide complete and reference designs, bills of materials, components and software/firmware to manufacture WiMAX-certifiable equipment.

For the manufacturers the total cost of chipsets will initially vary around a base price point of $100. The less-than-$100 price point is precisely what vendors are trying to achieve. The CPE chip players include both large players and small- and medium-sized start-ups. On one hand, a giant like Intel is interested in this market; on the other

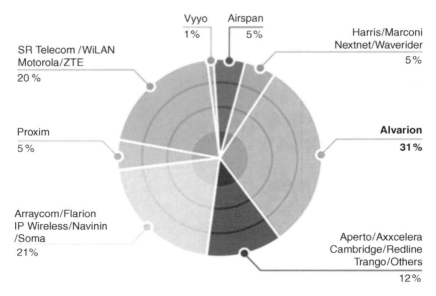

Figure 11.4 BWA vendor market share 2004

hand, start-ups like Cygnus Communications are working aggressively to take a substantial piece. However, although the big players have already released silicon products for WiMAX, start-ups seem to have more 'foil ware' than products (Figure 11.4).

11.2 ORIGINAL EQUIPMENT MANUFACTURERS

WiMAX equipment vendors come from two distinct camps, those historically involved with the fixed-wireless access business and the newcomers to the market. Companies that have, or are developing, WiMAX-compatible products for the fixed broadband access markets include:

- Airspan Networks Inc.;
- Alvarion Ltd;
- Aperto Networks;
- Cambridge Broadband Ltd;
- Navini Networks Inc.;
- Proxim Corp.;
- Redline Communications Inc.;

- Siemens AG;
- SR Telecom Inc.;
- Wi-LAN Inc.;
- ZTE Corp.

Note that, although these vendors are currently focused on 802.16-2004 'Fixed WiMAX', many of them are also plotting a migration to the newer 802.16e spec, which is more suitable for portable and mobile subscriber stations.

Newcomers to the market are typically targetting mobility and the newer 802.16e spec from the outset, and it is this that has attracted several of the mobile infrastructure market's big guns to WiMAX. Vendors that have announced initiatives include:

- Adaptix Inc.;
- Alcatel;
- LG Electronics Inc.;
- Motorola Inc.;
- Navini Networks Inc.;
- Nokia Corp.;
- Nortel Networks Ltd;
- Posdata Co. Ltd;
- Samsung Electronics Co. Ltd;
- Lucent Technologies Inc.;
- Ericsson AB.

Vendors will see that WiMAX and Mesh Network product market provide excellent opportunities. More multi-million dollar networking projects will be seen in next few years than ever. A major thrust will come from metropolitan markets, especially those run by municipalities.

On the operator side, more new players will emerge, with most of the buying in the next few years coming from players not even existing in the telecom operations landscape today, while other big segments will be held by non-3G cellular companies.

Amongst various types of equipment, the inclination will surely be heavily towards the unlicensed and mobile segments. The need for integrated infrastructure, that is, mobile (GSM/CDMA) and broadband, will rise steadily.

The key to success will be speed in delivering the new product. One of the key methods of accelerating uptake of a new technology is the development of reference platforms, which reduce time to market for

OEMs. Depending on the area of expertise, vendors must come out with such solutions as early as possible.

11.3   CHIP MANUFACTURERS

There are a number of start-ups and established semiconductor firms angling for a piece of WiMAX business. In the first phase, vendors are targeting the 802.16-2004 fixed WiMAX market with devices intended for outdoor CPE devices and self-install desktop modems. Chipsets for notebook PCs represent the second phase of development and could be available as soon as next year.

Developers are also working on devices for the upcoming 802.16e 'Mobile WiMAX' standard, with some start-ups focusing exclusively on this opportunity. Production of 802.16e chipsets is generally expected around the 2007–2008 timeframe.

Although a large volume of the WiMAX chipset market will be catering to the end-user device segment, there are also vendors focused on the base-station infrastructure, where greater processing power may be required and power consumption is less of an issue.

Here is an indicative list of WiMAX silicon vendors:

- Altera Corp.;
- Analog Devices Inc.;
- Aspex Semiconductor Ltd.;
- Beceem Communications Inc.;
- Cygnus Communications Inc.;
- Fujitsu Ltd.;
- Intel Corp.;
- PicoChip Designs Ltd.;
- Runcom Technologies Ltd.;
- Sequans Communications;
- TeleCIS Wireless Inc.;
- Wavesat Wireless Inc.;
- Xilinx Inc.

While most of the above-listed developers are working on 802.16 PHY and MAC components, or system-on-a-chip products, there are also a number of specialist suppliers of RF 'front-end' silicon for WiMAX devices. They include

- Texas Instruments Inc.;
- Atmel Corp.;
- SiGe Semiconductor Inc.;
- RF Magic Inc.

There are also firms focused on providing WiMAX software to chipset developers and equipment manufacturers. Examples include SiWave Corp. and Wi-LAN Inc. (Toronto: WIN – message board). In the smart antenna space, vendors such as ArrayComm Inc. and Roke Manor Research Ltd are also involved with WiMAX.

Companies noticeable by their absence in this segment include:

- Broadcom Corp.;
- Freescale Semiconductor Inc.;
- Philips Semiconductors;
- Qualcomm Inc.;
- STMicroelectronics NV.

11.4 DYNAMICS OF THE VALUE CHAIN

In the midst of this dramatic technological change, the industry is facing significant market-side challenges. Innovative new services are proving effective in attracting subscribers and enhancing subscriber loyalty, so there is intense competition to gain first-to-market advantages.

WiMAX vendors are trying to get their products into the market at an early stage, to prevent non-WiMAX BWA solutions from gaining too strong a foothold to be dislodged. Such non-WiMAX solutions are being driven by operators who need to be increase their revenues and offset the decline in conventional telephony and Internet services by moving into high speed, data-optimized mobility. This existing demand is a powerful attractive factor for the makers of broadband wireless equipment.

Until recently, mobility was the preserve of a few specialist vendors, many of whom have now shifted into the WiMAX camp and are promising migration paths to 802.16e. Most of the long-term WiMAX players are taking mobility very seriously as the greatest opportunity of their lives.

Also hotly pursued is the promising area of CDMA and GSM network upgrades to the future mobile 802.16e specification. 'Network-in-a-box', which collapses the functionality of all three elements (a softswitch, base station controller and mobile switching centre) elements into a single rack for GSM applications is providing a turnkey solution for carriers. This type of solution will be valuable in markets where budgets are tight, especially in rural or developing regions where populations may be sparse and carriers small.

11.5   RECOMMENDATIONS

Vendors: Reducing Costs and Accelerating Time-to-market

All companies can improve their competitive position in three key ways.

- accelerating their time-to-market for next-generation solutions;
- decreasing their development costs;
- improving the ability of their solutions to scale and adapt to evolving needs.

12

Strategy for Success: Government and Regulators

The governments and regulators should give a high priority to ensuring that people have access to broadband services through multiple facility-based platforms. Despite its relatively small share of the broadband market, wireless broadband has substantial potential for growth, as evidenced by the growing number of people who use wireless devices, such as cell phones or Wi-Fi-enabled laptops, to connect to the Internet.

Wireless services, especially wireless broadband, are critical to the evolution of communications infrastructure, and consequently to the health of a country's economy. Wireless is allowing users to access media, communications and computer resources and services more flexibly and is supporting the evolution of a diverse array of 'follow-me anywhere, available everywhere, always connected' functionalities.

Broadband wireless service has the potential to compete with wireline technologies in urban and suburban markets as a primary pipe to the home and business, to complement wireline technologies by adding a component of mobility or portability and to lead the way in rural markets where other broadband technologies are less feasible.

Taking the clue from this trend and considering the positives of wireless broadband access, authorities should demonstrate a strong commitment to facilitating wireless broadband investment and deployment, particularly through making it easier for entities to gain access to spectrum and employ new and advanced technologies that serve to provide wireless broadband to the public. They should take significant steps to facilitate the deployment of broadband wireless services.

The Business of WiMAX   Deepak Pareek
© 2006 John Wiley & Sons, Ltd

Regulators must aim to meet three general goals of increasing the availability of spectrum that can be used in the provision of broadband services, allowing maximum technical and regulatory flexibility for entities seeking to provide wireless broadband, and facilitating the development of the wireless broadband infrastructure by providing more regulatory certainty and removing regulatory disincentives.

Wireless is creating opportunities for wholly new services (e.g. distributed sensor nets, location-based services and mobile multimedia); it is enhancing the usability and lowering the costs associated with traditional computing and communications services (e.g. wireless drops to connect neighbourhood fibres to homes); and it is providing new platforms for competitive entry and new options for extending service to previously underserved communities (e.g. WISPs in rural communities).

While 'wireless services' encompass an incredible diversity of uses, technologies and markets, one commonality is that they all depend on access to the RF spectrum.

The traditional model for managing spectrum for commercial use has been based on a licensing regime that grants licensees limited and restrictive-use rights to a specific frequency band in a geographic area. There is general agreement among communication service providers, equipment makers, end-users and most policy-makers and industry analysts that these traditional mechanisms for managing RF spectrum are inefficient. This regime – predicated on 100-year-old radio technologies – is responsible for the substantial artificial spectrum scarcity and is badly in need of reform.

12.1 MAKING MORE SPECTRUM AVAILABLE

A crucial ingredient in the development of broadband applications and services over wireless networks is the availability of sufficient spectrum for the provision of wireless broadband. To that end, the regulator can take several steps to make more spectrum available for wireless broadband use for both unlicensed and licensed wireless broadband technologies.

Many, if not most, of the economists who have considered the issue appear to concur with the view that increased reliance on market forces would enhance efficiency and support assigning spectrum via transferable, flexible licences, especially when spectrum is perceived to be scarce.

12.2 IDEAL FRAMEWORK FOR SPECTRUM ALLOCATION

- Designate significant spectrum blocks for unlicensed use, meaning that anyone can transmit.
- Create an environment that facilitates transmitters in the new unlicensed bands cooperating intelligently for maximum efficiency.
- Explore innovative ways to improve and streamline the process of allocating and assigning licensed spectrum.
- Adopt more 'flexible use' policies and take additional steps to enable wireless broadband providers to use spectrum in licensed bands and to gain access to that spectrum.

Entrepreneurs should be allowed to build and operate wireless data networks based on 'open spectrum', much like how the cable operators did in the early 1990s. Over time, there was consolidation in the industry, as will happen in the wireless industry. Aggregators will also come in.

Wireless Broadband Services Using Networks of Unlicensed Devices

The regulator can adopt policies that can significantly foster growth in the provision of wireless broadband using unlicensed devices. More innovative wireless services must be promoted using unlicensed band. This can be achieved by setting flexible technical rules for operation of these devices as well as making available additional spectrum for these devices. Taken together, these policies foster continuous innovation in provision of wireless broadband services.

As seen in the past, two of the most positive developments for wireless broadband using unlicensed devices are the dramatic increase in the number of WISPs and the proliferation of Wi-Fi hot spots throughout the world. With off-the-shelf or readily available equipment and minimal investment, unlicensed WISPs are providing broadband connectivity to communities that previously had no broadband access and are also providing a competitive alternative to cable and DSL services. A decade ago, WISPs did not exist as an industry and now, depending on which estimate you take, there are more than 8000 WISPs in the USA alone, while there are 60 million Wi-Fi clients and 150 000 Wi-Fi hot spots globally.

Wi-Fi provides a different sort of broadband connectivity. Its ubiquity enables consumers to have access to broadband in many places outside the home or office – from airports to coffee houses to public parks.

The trend in the number of equipment authorizations is one indicator that shows how significantly this segment of the wireless broadband market has grown. As recently as 1995, there were approximately 1000 equipment authorizations granted annually. Last year, the number exceeded 2500, representing a 150 % increase over a span of less than a decade. While these data are useful indicators of the upward trend in the number of devices, these figures represent only the number of types of authorized devices and do not reflect the total number of devices deployed. Wireless networking devices of all kinds represent a significant number of total equipment authorizations. In addition to changing various Commission rules to facilitate wireless broadband, we have streamlined, and made more market-oriented the procedures associated with the equipment authorization process, significantly reducing the time-to-market for new wireless broadband products. For example, equipment manufacturers can now select from several private certification laboratories, in addition to the Commission's laboratory.

The continued growth in the use of unlicensed devices to provide broadband services is due to the fact that there are few barriers to entry in this market. Equipment costs are relatively low and equipment is available either off-the-shelf or readily from vendors of wireless networking equipment. In part, this reflects the relative success of the IEEE 802.11 family of standards and the excitement about the IEEE 802.16 family of standards; these ubiquitous, open standards ensure the interoperability of equipment and have effectively reduced the price point for wireless networking equipment.

Encourage Voluntary Frequency Coordination Efforts

As the radio spectrum is used more intensively, interference mitigation among unlicensed users is an increasingly important issue. The rules provide that unlicensed devices may not cause harmful interference to authorized users and must accept any interference that they receive. Moreover, unlicensed devices operating in a spectrum band do not have any preferred standing. Thus, as more and more devices use a particular unlicensed band in a localized area, interference mitigation will become increasingly important and, correspondingly, more technically complex.

Owing to the 'always-on' nature of broadband service, as compared with operations of other types of unlicensed devices with relatively shorter duty cycles, WISPs are more consistent and often more bandwidth-intensive users of spectrum. Thus, WISPs have even greater incentives to develop practices and procedures to mitigate interference.

Various voluntary private industry efforts are underway globally in which groups of unlicensed wireless service providers have set up databases and procedures to perform frequency coordination. These groups have found that these efforts substantially mitigate potential interference and facilitate quality of service. Authorities should support these private industry efforts if these frequency coordination initiatives encourage all spectrum users to become members. A key benefit of frequency coordination is the ability for more operators to share the same spectrum bands, avoiding the time-consuming, costly and often difficult task of determining the cause or source of any interference. Another principal benefit is enhanced service reliability.

In light of the benefits of frequency coordination groups and, given the continued growth in unlicensed wireless broadband services, more and more service providers will be interested in participating in frequency coordination efforts. To this end, we recently learned that the Licence Exempt Alliance is working to establish a nationwide frequency coordination database, which would serve as a referral in case of any problem.

One critical aspect for success of all of these private industry efforts is that they remain voluntary industry initiatives and regulators refrain from taking an active role in frequency coordination efforts in the unlicensed bands, as the industry members are in the best position to determine the optimal nature and extent of such coordination.

While increased growth of frequency coordination groups will be helpful in enabling more intensive use of the radio spectrum, voluntary industry 'best practices' will further facilitate this objective as well. For example, such practices could encourage the use of more spectrally efficient directional antennas and encourage service providers to transmit only when there is data to transmit.

Improving Access to Licensed Spectrum

Regulators must explore innovative ways to improve and streamline the process of allocating and assigning licensed spectrum. Although using licensed spectrum provides many advantages for wireless providers, one of the disadvantages is the lengthy period of time taken to allocate and assign new spectrum. Shortening the amount of time it takes to get spectrum out of the government's hands and into the market, where companies can use it to provide services that consumers demand, is critical in the fast-paced and ever-changing world of technology and broadband.

Regulators must continue to explore new ways to reduce the amount of time between allocation and assignment, for example by simultaneously allocating and proposing service rules for spectrum, as recently done by the FCC in the USA in the case of Advanced Wireless Services.

Furthermore, in cases where parties disagree on the appropriate band plan for a new spectrum block, the regulator could consider resolving technical disputes over allocation schemes at auction by using competitive bidding to determine the band plan most highly valued by prospective licensees and then move forward with licensing based on the winning band plan.

The regulators should try to allocate spectrum that is in harmony with international spectrum allocations. The use of a single band for the same service across multiple countries can create economies of scale in the production of wireless end-user equipment. This in turn can lower the cost of broadband-capable devices, thereby increasing the demand for broadband services and making them more accessible to a wider base of consumers. Global harmonization can also facilitate international roaming, which can increase the productivity of workers who use broadband devices when travelling around the world.

Increasing Technical and Regulatory Flexibility for Licensed Band

The regulator must promote the efficient use of spectrum by giving licensees the flexibility to choose which technologies and services to deploy using the spectrum they hold. Under this spectrum management model, licensees can deploy the technologies or services that best fit their business plans and that meet the demands of their customers, as long as doing so complies with the technical requirements of the license and does not cause interference to with adjacent licensees.

Owing to the growing demand for spectrum that can be used for new and emerging technologies, it has been increasingly important for spectrum allocations and subsequent service rules to be flexible and designed to facilitate as many types of offerings as possible.

As compared with other alternatives, the general adoption of a more flexible and market-oriented approach to spectrum policy is the better course to provide incentives for users to migrate to more technologically innovative and economically efficient use of the spectrum, and to provide the services that markets determine are most valued, including broadband services.

The regulator should consider developing innovative approaches to enable incumbent licensees to obtain additional flexibility that would facilitate provision of wireless broadband and other advanced services.

One approach would be to consider granting additional flexibility to incumbent licensees through significant revisions of the applicable service rules. Alternatively, the regulator can consider various possible market-based auction mechanisms that could be used to provide additional flexibility to incumbent licensees. For instance, the Commission could consider employing mechanisms whereby spectrum previously licensed to incumbent licensees would be made available at auction with different rights (e.g. flexible use), and potentially could be combined with other spectrum, including that not previously licensed (e.g. 'white space'). Such mechanisms could give incumbent licensees the option to return their current licenses in exchange for the means to obtain comparable spectrum access. In this regard, we recommend that the Commission explore various methods by which this framework might be implemented. These include providing an auction in which incumbents would exchange their licenses for tradable bidding offset credits, the value of which would be linked to the winning bids for licenses sold in the auction. Another possible option would be to conduct an auction which permits incumbents to participate not only as potential buyers but also as sellers of their existing licenses, with the right to set a reserve price below which they would choose not to sell the licenses.

12.3   REDUCING LEGACY REGULATION

Apply a Deregulatory Framework – One that Minimizes Regulatory Barriers to Wireless Broadband Services

Imposing unnecessary and conflicting regulatory requirements on wireless broadband providers impedes the deployment of these services. In the absence of clear Federal guidance, a number of local authorities have begun to regulate these services, resulting in additional costs to the providers and, ultimately, to consumers.

Specifically, the prospect of inconsistent and burdensome regulations threatens to hinder investment in, and delay deployment of, wireless broadband services. Adoption of varying and inconsistent state and local regulation harms consumer welfare by reducing the economic efficiencies inherent in a national market.

Regulators must adopt policies that establish a deregulatory framework for wireless broadband and adhere to the following general, overarching principles:

- Minimize regulatory barriers at the federal level through a deregulatory approach and eliminate unnecessary federal regulatory barriers that impede the development of wireless broadband services. This will allow market incentives to bring about rapid and ubiquitous broadband deployment and innovation. To the extent possible, it should be ensured that all types of wireless broadband – mobile, portable and fixed – are regulated in a similar manner. Given the rapidly evolving and innovative nature of the broadband services generally, regulators should let the marketplace direct the development of services over wireless broadband rather than risk hindering its growth through regulation.

- Take a pro-competitive, pro-innovative market-based approach. Equally importantly, regulators should not attempt to pace the technological advancements and changes in consumer preferences with its rules, but should instead allow the market to determine the development and resulting deployment of broadband services.

- Adopt a framework that prevents inconsistent regulation and minimizes regulatory requirements at the state level. Extensive regulation at the state level could create certain disincentives to deploy broadband facilities. Inconsistent state regulations could delay the provision of the service in some areas of the country, impacting the services offered even outside the heavily regulated states. Further, burdensome regulations can inhibit innovation in wireless broadband services and deny national providers the ability to achieve the benefits of economies of scale that their products need to succeed.

- Creating conditions for competitive economics. The last 20 years have shown convincingly the power of markets to deliver services and good value-for-money in the telecommunications services industry. It is essential, therefore, that the market be not ignored.

Non-commercial and interventionist approaches should be undertaken only where necessary, and even then non-commercial provision should exploit markets where possible.

12.4 GOVERNMENTS CAN MAKE A POSITIVE IMPACT

BWA technologies and their applications are in their early stages of development, and their potential economic and social benefits appear to be considerable. These new technologies and applications, however, are emerging in many different situations, often outside the operational landscape of traditional telecommunications services, and with new types of participants from both the private and public sectors.

BWA initiatives may be seen as disruptive and can hit unintended roadblocks in the form of local regulations and lack of understanding of their potential. Conversely, proper government support and incentives can accelerate their successful implementation at little cost and with significant immediate economic and social benefits for the poor.

Each country presents unique characteristics and conditions with respect to BWA deployment, from a geographic, social, economic, regulatory and telecommunications infrastructure standpoints. There are, however, several regulatory and economic factors that regulatory authorities should keep in mind while performing their duties. We will discuss some of the key issues which have a direct bearing on the success of BWA initiatives as the well as the resulting sustained balanced economic development.

Encourage the Aggregation of Demand for Bandwidth

One of the most important factors of success for wireless ISPs is a rapid increase in initial demand for connectivity, which allows for a faster break-even on operating expenses. This can be achieved through initial aggregation of demand based on applications for local public services such as schools, universities, health services and public administration. Business, agriculture and private use will inevitably add to the mix once service is available. In underserved areas, it is likely that initial viable aggregation will occur through the deployment of wireless Internet kiosks operated by entrepreneurs.

Promote BWA Technologies

Identify, promote and establish national consensus on the potential benefits of BWA technologies. The success of any public technology initiative, including BWA deployment, depends heavily on a generally supportive environment, a prerequisite of which is that policy makers, the public and private sectors and the local media have adequate awareness about benefits and other key issues relevant to the technology in question. Raising awareness and building national consensus about the benefits of low-cost broadband wireless Internet infrastructure solutions is therefore an important step.

Identifying leading applications, which may drive the initial use of wireless Internet infrastructure and distribution, will further develop support for wireless Internet solutions among key constituents. Governments should also encourage local public services to use the infrastructure of local wireless ISPs. Local governments may also inventory those applications that may contribute most to bridging the digital divide both from a geographic and a social standpoint, and foster economic development, job creation and productivity gains in all economic sectors.

Create an Environment for Collaborations

Create an environment for collaborations at governmental and intergovernmental levels, including sharing best practices. It is anticipated that wireless technologies and applications will continue to evolve rapidly, which makes it important for governments and private sector leaders to remain abreast of other countries' experiences, regulatory work at the international level, best practices and latest innovations. Governments must encourage knowledge sharing amongst their own constituents as well as with other countries to leverage the existing knowledge and experiences of other organizations. Sharing of ideas can be very productive, especially in the area of e-government, e-education and e-health.

Wireless Internet infrastructure and services can leverage a number of existing resources in any given country. Some of the areas and players which can be considered for fostering cooperation are:

- backbone (operators and owners of fibre-optic networks including governments, private sector networks, telecommunication companies, power-grid operators and satellite communications operators);

- location (owners of land and high points with adequate power supply and security to instal antennas, such as existing radio communications towers or possibly public-sector buildings such as post offices or other types of standard venues);
- expertise (systems integrators with the technical capabilities to instal and maintain wireless equipment such as towers, cabling, hub-integrated antennas, wireless modems, control and network management systems, routers, cables, uninterruptible power supplies, racks, etc.);
- know-how (operators of similar services such as TV broadcasting, cellular telephony, computer maintenance organizations and power distribution);
- funds (financial sector interested to fund start-up ISPs).

Such initiatives can lead to the shortening of project cycles as well as reducing cost, and hence better chances of success.

12.5 RECOMMENDATIONS

Regulators – deregulate as much as possible, let the market find the best regulatory environment.

Governments – promote the benefits of BWA and create an environment for collaboration and cooperation.

13

Strategy for Success: Users and Investors

13.1 WIRELESS – INVESTMENT PICKS UP

Wireless was among the sectors most affected when the high-tech bubble burst in 2001. The percentage of venture capital (VC) that the wireless industry received dropped from 22 % in 2000 to only 9 % in 2001 and 10 % in 2002. Since then investment in wireless industry has started to improv. Now the wireless industry has become a bright spot in high-tech industry investment, both in equipment and in services.

Investment by VCs in the wireless industry increased to $1.077 billion in the first quarter of 2003 from $658 million in the last quarter of 2002. The investment level in wireless has been consistent for the past seven quarters at an average of $918 million per quarter. It is estimated that the wireless industry now receives 13–19 % of all venture capital investment.

Where the Dollar is Going

Five per cent of wireless investment goes to early-stage companies, 38 % to middle stage and 57 % to later stage companies. Venture capital interests reflect the future directions of the wireless sector.

Currently, most VC investment in wireless goes to satellite communications, broadband wireless, mobile data acceleration (MDA), RF semiconductors, directory assistance software (DA), ultra wideband (UWB), ZigBee and protocol software.

The Business of WiMAX   Edited by Deepak Pareek
© 2006 John Wiley & Sons, Ltd

Some other areas in wireless have also caught attention of investors, including fixed wireless broadband, network management software, power management chips, directory assistance software and protocol software.

Investment and Sustainability

The success of the wireless sector (and prevention of another 'bubble') depends not only on the success of companies in attracting investment but also on the sustainability of their business models. The dot.com bubble was driven by the most exuberant IPO market in history from the beginning of 1997 through to the third quarter of 1999.

Technology companies had easy access to inexpensive capital in the IPO market. The average time for VC development, management and financing of a portfolio company before going public was shortened to only 6 months to 3 years, compared with an average of 4–8 years before 1997. However, the dot.com boom became a pyramid scheme when some investors realized that they could get an exponential return by cashing out after an IPO was launched because stock market investors were rushing to buy dot.com stocks and pushing market values to unrealistic levels. VCs were lured to pay less attention to sustainable business because the market accepted immature business models. When the IPO money was gone, many companies fell apart.

Could this happen again with wireless? In the era of easy IPO money, the valuation of portfolio companies in pre-IPO rounds increased dramatically. The percentage increase of the pre-IPO rounds was significantly higher than in previous financing rounds, implying that most of the 'bubble' in investment valuation came from the pre-IPO rounds financing.

To avoid another bubble, the IPO market needs to be watched closely. Currently, the IPO market is one of the worst for venture-backed companies in two decades. *Entrepreneur* magazine stated that, in 2002, venture capital investing fell to its lowest level in 5 years ($21.2 billion). There are less than 50 venture-backed IPOs per year, and later-stage companies are usually significantly undervalued. On the other hand, the merger and acquisition (M&A) market for venture-backed companies has remained strong since 2001.

Unlike the IPO market, typical buyers in M&A transactions are usually large companies seeking expansion or entering a new market. They are more knowledgeable in the specific field and much more cautious about sustainability. Given the current situation in the capital market, another bubble in technology companies is unlikely, at least in the short term. However, with the economy recovering, and the capital market rebounding, the venture-backed technology companies may once again have easy access to cheap capital.

Past lessons are key to seeing danger.

13.2   BWA: INDUSTRY MATURES

In the past 3 years BWA has received more than $1.6 billion in VC investment worldwide. In the first quarter of 2003 alone, 22 BWA companies received $70 million. Although this is the area where venture capitalists see huge potential, there is a growing consensus in the VC industry that there is a great deal of disparity and randomness in BWA sector funding; for example, Wi-Fi was overfunded while some sectors like WiMAX were underfunded.

VC targets in this space have been semiconductor developers, providers of platforms for managing access, system technology developers and network operations companies. In the past two years, BWA investors have shifted their focus from semiconductor companies to systems and operations companies. Network operations companies, including roaming billing, administration, efficiency and security, have received 34 % of total VC funding compared with 31 % for chipset companies.

13.3   WiMAX: BUBBLE OR LIFETIME OPPORTUNITY

The evolving world of WiMAX gained a new aura of respectability when silicon kingmaker Intel threw its weight behind the prestandard broadband wireless technology in 2003 and its investor arm, Intel Capital, pledged $150 million to help nurse the fledgling sector, which was started last year. However, despite Intel's considerable influence, investors and venture capitalists are largely reluctant to follow in Intel's big footprints and invest in WiMAX, according to analysts.

The current level of investor interest in WiMAX is fairly high but there is some trepidation among the pure VCs because of the bad associations with previous broadband wireless failures like Teligent. Part of investors' hesitancy stems from the sheer complexity of the business – the need for spectrum and standards and the competing wireless broadband technologies such as EV-DO, HSDPA and Wi-Fi.

However, the eye-rolling may also be partly a function of the inherent impatience of the VC mind, stimulated by Intel's cheerleading and frustrated by the reality that mobile WiMAX is not expected until 2008.

There are two schools of thoughts, or two camps of VCs: one camp simply disagrees that WiMAX is as promising as Intel claims. The other not only sees the value of WiMAX but also believes the opportunity it brings really belongs to large companies rather than start-ups, essentially leaving VCs out of the game.

Not all VCs feel that way, evidently. Those open to WiMAX seem particularly interested in WiMAX chips, applications and roaming software. California chip start-up TeleCIS roughly doubled its total funding in December with a $4 million B round from ATA Ventures. Canadian chip vendor Wavesat closed an oversubscribed $10.5 million round of VC funding last fall, including $2.5 million from a pension fund, and in February, French chip start-up Sequans Communications announced an oversubscribed $9 million B round from European sources.

The Hurdles

The immense economic and social potential inherent in WiMAX notwithstanding, the industry overall still faces significant challenges:

- business models for emerging wireless companies are unclear;
- immature industry standards result in solutions that are not quite ready for the mass market – WiMAX mobile is still in the standards development stage;
- most of the expectation built by vendors is on promises, not reality.

Investor confidence and overall perception of WiMAX have suffered recently due to the delay of about 6 months in the 802.16-2004 certification process. Like an overexcited child, the technology industry is never able to stop itself wishing away the days until Christmas. Time and again, standards and products fail to meet their schedules, not because of any intrinsic problem, but because the deadlines were unrealistic to start with.

The delay is a blow that is more symbolic than real. However, the delay will prompt scepticism about the program, and this will be far more serious if it has a knock-on effect on the upcoming mobile standard, 802.16e. It is essential that, from now on, the WiMAX

Forum sets realistic deadlines and does not risk further backlash against its technologies.

Another important thing to look out for is that such roadblocks, however, small, can enhance investors' hesitancy, part of which stems from previous broadband wireless failures like Teligent, and the rest from the sheer complexity of the WiMAX business.

These are legitimate, immediate concerns for the wireless industry as a whole; resolving them does not put the region at a specific competitive disadvantage within the industry. However, enthusiasm for the industry must be tempered by a consideration of these challenges.

On the other hand, their existence, while daunting, may also point to new business opportunities, for example, in standards integration and the development of business strategies. It is like being in the Wild West, unsettled but filled with opportunities for those who persevere.

Prospect Areas

After a long time investors are finding some solace as well as appeal in the telecommunications sector. For WiMAX to work economically (one of the main sticking points is the need to lower the cost of its CPE from more than $200 to less than $100), it must be manufactured and deployed in large volumes by big vendors such as Alcatel, Lucent Technologies, Siemens and Nortel Networks – or so the theory goes – which is why VCs feel left out.

However, the applications that will ride on future WiMAX networks (interactive games, for example) could well be developed by start-ups, which is why VCs could find an important role at that level so long as they can wait for the networks to be built first. With a confusing array of competing wireless broadband technologies such as EV-DO, HSDPA and Wi-Fi, investors are growing increasingly interested in chips and software that might enable handoffs between WiMAX and other forms of wireless broadband.

13.4 RECOMMENDATIONS

Investors: caution is the buzz word.

After a long time investors are finding some solace as well as appeal in the telecommunications sector. One strong point in favour of WiMAX is that wireless economics have already been proved correct in mobile voice services, where the cost per line is 40 % of the cost of equivalent wireline services. These same economies are expected to hold true in wireless data using WiMAX.

Some facts which can predict the future success of WiMAX, or at least confirm the march in right direction, is that the number of members in the WiMAX Forum has grown from less than 50 to 250 in the last year and the number of operator members has grown from 3 to 80.

APPENDICES

A1

WiMAX Certification Process

The combination of WiMAX conformance and interoperability testing makes up what is commonly referred to as certification testing. WiMAX conformance testing can be done by either the certification laboratory or another test laboratory and is a process where base station and subscriber station manufacturers test their pre-production or produced units to ensure that they perform in accordance with the specifications called out in the WiMAX Protocol Implementation Conformance Specification (PICS) document.

Based on the results of conformance testing, BS/SS vendors may choose to modify their hardware and/or firmware and formally resubmit these units for conformance testing. The conformance testing process may be subject to a vendor's personal interpretation of the IEEE standard, but the BS/SS units must pass all mandatory and prohibited test conditions called for by the test plan for a specific system profile.

WiMAX interoperability is a multivendor (≥ 3) test process hosted by the certification laboratory to test the performance of BS and/or SS in transmitting from one vendor and receiving data bursts from another based on the WiMAX PICS. Figure A1.1 shows the preliminary WiMAX certification process with its components.

First, the vendor submits BS/SS to the certification laboratory for Pre-certification qualification testing where a subset of the WiMAX conformance and interoperability test cases is done. These test results are used to determine if the vendor's products are ready to start the formal WiMAX conformance testing process.

Upon successful completion of the conformance testing, the certification laboratory can start full interoperability testing. However, if the

The Business of WiMAX Deepak Pareek
© 2006 John Wiley & Sons, Ltd

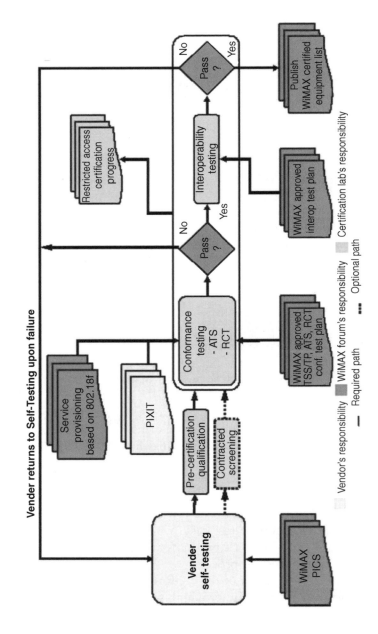

Figure A1.1 Certification process

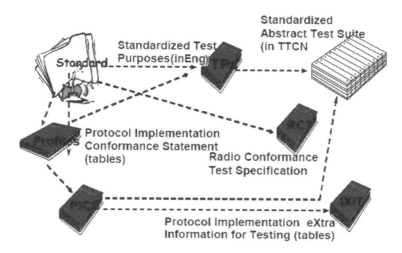

Figure A1.2   Abstract test suite development process

vendor's BS/SS failed some of the test cases, the vendor must first fix or make the necessary changes to their products (BS, SS) and provide the upgraded BS/SS with the self-test results to the certification laboratory before additional conformance and regulatory testing can be done.

If the BS/SS fails the interoperability testing, the vendor must make the necessary firmware/software modifications and then resubmit their products with the self-test results for a partial conformance testing, depending on the type of failure and the required modification.

The end goal is to show service providers and end-users that, as WiMAX Forum Certified hardware becomes available, service providers will have the option of mixing and matching different BSs and SSs from different vendors in their deployed networks.

Upon successful completion of the described process flow, the WiMAX Forum will then grant and publish a vendor's product as WiMAX Forum Certified. It should be pointed out that each BS/SS must also pass regulatory testing, which is an independent parallel process to the WiMAX certification process (Figure A1.2).

ABSTRACT TEST SUITE PROCESS

The WiMAX Forum is working on the development of numerous process and procedural test documents under the umbrella of the

IEEE 802.16 standard. The key WiMAX test documents are as follows:

- PICS in a table format;
- test purposes and test suite structure (TP and TSS);
- radio conformance test specification (RCT);
- protocol implementation extra information for testing (IXIT) in a table format.

These test documents are used in the development of a standardized abstract test suite (ATS). The ATS is the culmination of test scripts written in a tree and tabular combined notation (TTCN) language. The end products of the ATS are test scripts for conformance and interoperability testing under a number of test conditions called for in the PICS document for a specified WiMAX system profile. Test scripts automate the process of WiMAX certification testing.

A2

WiBro

One development worth mentioning here is WiBro (wireless broadband). WiBro is a South Korean initiative and an opportunity for the country to establish a 'homegrown' wireless technology, much similar to what the Chinese are doing with TD-SCDMA (Figure A2.1).

WiBro will now probably be included within the .16e umbrella, thus making it another potential WiMAX profile. Specifically, WiBro is a

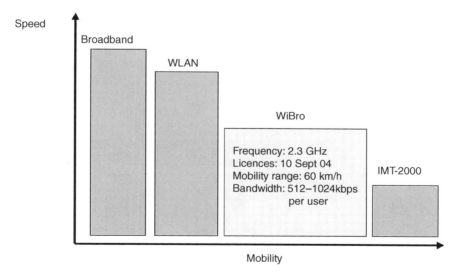

Figure A2.1   WiBro speed vs mobility

The Business of WiMAX   Deepak Pareek
© 2006 John Wiley & Sons, Ltd

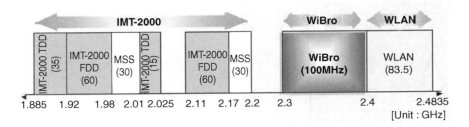

Figure A2.2 WiBro spectrum

TDD-based system that operates in a 9 MHz radio channel at 2.3 GHz with OFDMA as its access technology (Figure A2.2).

According to its proponents, WiBro supports users travelling at speeds up to 120 km/h (previously it was advertised as being limited to 60 km/h), with peak user data rates of 3 Mbps in the downlink (uplink = 1 Mbps) and 18 Mbps of peak sector throughput in the downlink (uplink = 6 Mbps). Average user data rates are advertised as being in excess of 512 kbps, and with the cell radius limited to 1 km, it will largely be deployed in densely populated areas (Figure A2.3).

Initially, WiBro was perceived as being a portable solution, even though it could support mobile users, since the technology did not

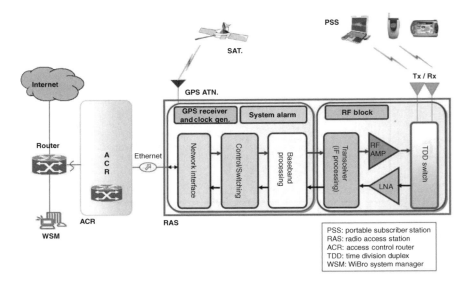

Figure A2.3 WiBro functional model

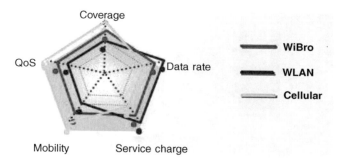

Figure A2.4 WiBro value analysis

support seamless cell handoffs. With its potential future adoption into the WiMAX family of profiles, there could be a desire to introduce vehicular mobility or near-seamless handoffs.

It is not entirely clear how WiMAX/WiBro will evolve, but it is foreseeable that the technology will first try to incorporate limited portable features, and then, based on customer demand, technology advancements and the underlying economics of an inherently more expensive solution, it will move towards more 'seamless mobility' (Figure A2.4).

Korea Telecom, in conjunction with Samsung, is promising commercial WiBro services by April 2006, while the Nortel and LG WiBro joint-venture, which was announced in March 2005, is currently suggesting customer trials in the second half of 2006. Given the lag between infrastructure and CPEs, the 'commercial' service will probably lack commercially viable CPEs until at least late 2006.

A3

Proprietary BWA Systems

Early commercial development of BWA was led by four major vendors

- Arraycomm;
- Flarion;
- IP Wireless;
- Navini.

Each has developed their own unique approach to achieving high speed broadband, leading to tradeoffs in operating spectrum requirements, coverage and cost.

ARRAYCOM

iBurst technology from Arraycom has features like:

- high broadband data speeds – up to 1.0 Mbps (downlink) per user;
- wide area coverage – its range is one of the best in the industry;
- always-on connectivity – the network supports full handoff;
- low cost – it claims to have a market-leading cost structure;
- simplicity – easy to deploy, easy to install, easy to use;
- commercialization – operationally proven and scalable;
- capacity – each base station sector can deliver more than 30 Mbps;
- use of TDD to permit downlink and uplink paths to share common spectrum;

The Business of WiMAX   Deepak Pareek
© 2006 John Wiley & Sons, Ltd

- less spectrum needed; however, it requires a regulator to make it available;
- use of the unique 'Intellicell' smart antenna spatial processing system;
- enhanced signal path between base station and customers.

FLARION

Flarion's technology has been tested at various locations and it has features like:

- high throughput – up to 1.5 Mbps (downlink) per user (Flexband claims 2.5 Mbps sustainable sector throughput with up to 800 kbps at cell edge);
- spectral efficiency (1.25 or 5 MHz multi-carrier);
- full mobility with handoff;
- low latency – less than 50 ms claimed;
- IP-based QoS system for the air interface;
- based on the use of Flash OFDM technology;
- available in frequency bands from 450 MHz to 3.5 GHz; however, the only deployments to date have been at 850 and 1900 MHz plus a trial at 700 MHz;
- cellular-like handset terminal with full VoIP functionality announced.

IP WIRELESS

IP wireless technology is one of the most successful of the proprietary BWA systems with features like:

- UMTS TDD standard (Release 99);
- moderate throughput rates – speeds can approach DSL performance;
- high-performance QoS scheme;
- compact base station systems;
- mature series of CPE devices;
- portable operation – handoff support is limited;
- scalable network architecture;
- plans to add HSDPA will increase capacity per cell by about 20%;
- variety of frequency bands in both TDD and FDD configurations between 1.9 and 3.5 GHz.

NAVINI

This technology, called 'Ripwave' System, has features like:

- high broadband data speeds – up to 6 Mbps (downlink) per user claimed;
- high capacity – each base station can deliver more than 12 Mbps;
- wide area coverage – high 163 dB link budget;
- low cost – claims up to 50 % lower than DSL or cable;
- simplicity – easy to deploy, easy to install, easy to use;
- low latency – 50–60 ms claimed;
- VoIP – demonstrated solution for broadband telephony;
- smart antenna system with phased antenna elements to improve coverage and signal quality;
- available in spectrum assignments between 2.3 and 3.5 GHz.

CONCLUSION

WiMAX is very much a global initiative. The original air interface standard (802.16) addressed applications in licensed bands (10–66 GHz). Subsequent amendments extended 802.16 to cover NLOS applications in licensed and unlicensed bands (sub-11 GHz). We now tend to categorize WiMAX applications into four versions or generations: fixed, nomadic, portable and mobile, the former being more complete IEEE standards than the latter.

When looking at WiMAX, it is important to view it as two distinct stages of evolution. The first stage will begin next year with products that cost and function much like current BWA equipment. The total fixed wireless market will not expand as a result of WiMAX; what we will see is a gradual migration of purchasing behaviour from proprietary equipment to WiMAX equipment. Operators will be wary of adopting WiMAX equipment until prices drop to the point where they cannot afford to ignore WiMAX, which should occur in late 2005.

At about the same time, we will see the beginning of the second stage of WiMAX: the birth of metro-area portability. Once 802.16e is approved, laptops and other mobile devices may be embedded with WiMAX chipsets, so that the users can have Internet access anywhere within WiMAX zones. If this sounds like 3G, in many ways it is. The second stage of WiMAX could be very disruptive to 3G operators and could drive a round of WiMAX network overlays in urban areas.

Nevertheless, this will not happen until 2006 at the earliest. As shown in the following exhibit, WiMAX (stages one and two) and WiFi will complement one another.

In its future mobile version, a WiMAX-enabled device will maintain its data session by concurrently connecting to multiple WiMAX base stations, but if some infrastructures (or devices) are not standards-compliant, there is the risk of dropping a user and corrupting their application data. With the WiMAX standards still evolving, with no certifications yet issued to give buyers confidence about standards-compliance, and with many unproven business models, some operators feel confused and hesitant about WiMAX roll-outs.

Consequently, the IEEE and ETSI have accelerated their standards-making in response to market demand. The nomadic standard (802.16d is now called 802.16-2004) was published in July 2004 to consolidate all amendments and base standards.

The mobile standard (802.16e) has reached a final draft, incorporating scalable signal modulation modes (SOFDMA) for the mobility standard. An *ad hoc* group has been tasked to enable roaming across networks, and ETSI HIPERMAN has been harmonized with 802.16-2004 OFDM.

WiMAX is emerging as a last-mile broadband wireless Internet access solution. WiMAX provides wireless services in the MAN just as Wi-Fi provides wireless services in LANs. WiMAX has the potential to make broadband service available in regions where it is currently not feasible, particularly in rural communities.

When certified products become available, the market will expand. Costs should be less than those for Wi-Fi because WiMAX is a standards-based technology. Spending on WiMAX infrastructure is expected to increase dramatically in the next few years, growing from $15 million in 2004 to $290 million by 2008, growing at a 109.7 % compound annual growth rate.

Infrastructure revenue includes CPE, point-to-point equipment used in backhauling LANs to the Internet and point-to-multipoint equipment used in broadband access. WiMAX is potentially disruptive in that it could compete with other high-speed fixed solutions, including DSL and cable modems, as well high-speed mobile solutions like 3G.

Glossary

1xEV-DO (1 × evolution-data optimized)	A technology that offers near-broadband packet data speeds for wireless access to the Internet. It is an alternative to wideband CDMA (WCDMA). Both are considered 3G technologies. A well-engineered 1xEV-DO network delivers average download data rates between 600 and 1200 kbps during off-peak hours, and between 150 and 300 kbps during peak hours. Instantaneous data rates are as high as 2.4 Mbps. Only 1.25 MHz of spectrum is required, one-quarter of that for WCDMA.
3G	Third-generation wireless, specified by the ITU promises to offer increased bandwidth and high-speed data applications up to 2 Mbps. It works over wireless–air interfaces such as GSM, TDMA and CDMA. 3G refers to near-future developments in personal and business wireless technology, especially mobile communications. This phase is expected to reach maturity between 2003 and 2005.
802.11	A family of specifications for wireless local area networks (WLANs) developed by a working group of the Institute of Electrical and Electronics Engineers

(IEEE). There are currently three specifications in the family, 802.11a, 802.11b and 802.11g. All use the Ethernet protocol and CSMA/CA (carrier sense multiple access with collision avoidance) for path sharing. 802.11e, 802.11h and 802.11i are in different stages of development and approval. This series of wireless standards developed by the IEEE is also commonly known as Wi-Fi.

802.11a   A wireless networking specification, assigned by the IEEE, in the 5 GHz frequency range with a bandwidth of 54 Mbps.

802.11b   A wireless networking specification, assigned by the IEEE, in the 2.4 GHz frequency range with a bandwidth of 11 Mbps.

802.11g   A wireless networking specification, assigned by the IEEE, in the 2.4 GHz frequency range with a bandwidth of 54 Mbps.

802.16   A group of broadband wireless communications standards for metropolitan area networks developed by a working group of the IEEE. 802.16 is the IEEE Air Interface Standard specification for fixed broadband wireless access systems (wireless metropolitan area networks, see MAN) employing a point-to-multipoint architecture.

802.20   A specification of physical and medium access control layers of an air interface for interoperable mobile broadband wireless access systems, operating in licensed bands below 3.5 GHz, optimized for IP-data transport, with peak data rates per user in excess of 1 Mbps. It supports various vehicular mobility classes up to 250 km/h in a MAN environment and targets spectral efficiencies, sustained user data rates and numbers of active users that are all significantly higher than those achieved by existing

	mobile systems. This standard is under development.
AES (advanced encryption standard)	An encryption algorithm for securing sensitive but unclassified material from US Government agencies and, as a likely consequence, may eventually become the *de facto* encryption standard for commercial transactions in the private sector.
Access point	A wireless hardware device connected to a wired network that enables wireless devices to connect to a wired LAN.
Analogue	Modulated radio signals that enable transfer of information such as voice and data.
Antenna	A device used for transmitting and/or receiving radio signals, whose shape and size is determined by the frequency of the signal it is receiving.
Authentication	A process of identifying a user, usually based on a username and password, ensuring that the individual is who he or she claims to be, without saying anything about the access rights of the individual.
Backbone	The central part of a large network to which two or more sub networks link. It is the primary path for data transmission. A network can have a wired backbone or a wireless backbone.
Bandwidth	The amount of data a network can carry, i.e. how much and how fast data flows on a given transmission path. It is measured in bits or bytes per second.
Base station	The central radio transmitter/receiver that maintains communications with mobile radiotelephone sets within a given range (typically a cell site).
Bits per second (bps)	Is the number of bits that can be sent or received per second over a communication line.
Bluetooth Wireless Technology	A short-range wireless specification that allows for radio connections (2.4 GHz) transmitting voice and data between devices (such as portable computers, personal digital assistants, PDAs, and

	mobile phones) within a 30-foot range of each other. Bluetooth is a computing and telecommunications industry specification that describes how mobile phones, computers, and PDAs can easily interconnect with each other and with home and business phones and computers using a short-range wireless connection.
Broadband	A fast Internet connection generally above 200 kbps. However, no official speed definition exists for broadband services.
CDMA (code division multiple access)	A form of multiplexing which allows numerous signals to occupy a single transmission channel, optimizing the use of available bandwidth. CDMA2000, also known as IMT-CDMA Multi-Carrier or IS-136, is a CDMA version of the IMT-2000 standard developed by the International Telecommunication Union (ITU). The CDMA2000 standard is third-generation (3G) mobile wireless technology. CDMA2000 can support mobile data communications at speeds ranging from 144 kbps to 2 Mbps.
Cable	A broadband transmission technology using coaxial cable or fibre-optic lines that was first used for TV and is now being used for Internet access.
Channel	A path along which a communications signal is transmitted.
Client device	That which communicates with the hub, e.g. access points and gateways. They include PC cards that slide into laptop computers, PCMCIA modules embedded in laptop computers, and mobile computing devices.
Client	An end user, i.e. any computer connected to a network that requests services (files, print capability) from another member of the network.
Consumer premises equipment (CPE)	Devices located at home or in the office such as telephones, PBXs, and other communication devices.

DES
(data encryption standard)

A widely used method of data encryption using a private (secret) key that was judged so difficult to break by the US Government that it was restricted for exportation to other countries. There are 72 quadrillion or more possible encryption keys that can be used. For each given message, the key is chosen at random from among this enormous number of keys. Like other private key cryptographic methods, both the sender and the receiver must know and use the same private key.

DNS

A program that accesses a database on a collection of Internet servers to translate URLs to Internet packet (IP) addresses.

DSL
(digital subscribers line)

Various technology protocols for high-speed data, voice and video transmission over ordinary twisted-pair copper POTS (plain old telephone service) telephone wires.

Dial-up

A communication connection using standard copper wire telephone network.

Digital

A technology used in telecommunications where information is processed by first converting it to a stream of ones and zeros, permitting extremely complicated systems to be designed and manufactured at reasonable cost through the use of application-specific ICs and computer circuitry while meeting very high performance standards.

E911
(short for enhanced 911)

A location technology advanced by the FCC that will enable mobile or cellular phones to process 911 emergency calls and enable emergency services to locate the geographic position of the caller.

EDGE
(enhanced data GSM environment)

A faster version of the global system for mobile (GSM) wireless service, designed to deliver data at rates up to 384 kbps and enable the delivery of multimedia and other broadband applications to mobile phone and computer users. The

EDGE standard is built on the existing GSM standard, using the same time-division multiple access (TDMA) frame structure and existing cell arrangements. The pioneer in GSM and DECT standards.

ETSI
(European Telecommunications Standardization Institute)
Ethernet (also called 10Base T)

Ethernet is an international standard for wired networks. It can offer a bandwidth of about 10 Mbps and up to 100 Mbps.

Flash-OFDM
(orthogonal frequency division multiplexing)

A new signal processing scheme from Lucent/Flarion that will support high data rates at very low packet and delay losses, also known as latencies, over a distributed all-IP wireless network. The low-latency will enable real-time mobile interactive and multimedia applications. It promises to deliver higher quality wireless service and better cost effectiveness than current wireless data technologies.

Firewall

Software, hardware or a combination of the two that prevents unrestricted access into or out of a network.

GERAN

A term used to describe a GSM- and EDGE-based 200 kHz radio access network. The GERAN is based on GSM/EDGE Release 99, and covers all new features for GSM Release 2000 and subsequent releases, with full backward compatibility to previous releases.

GPRS
(general packet radio services)

A packet-based wireless service that promises data rates from 56 up to 114 kbps and continuous connection to the Internet for mobile phone and computer users. The higher data rates will allow users to take part in video conferences and interact with multimedia Web sites and similar applications using mobile handheld devices as well as notebook computers. GPRS is based on global system for mobile (GSM) communication and will complement existing services such

GMSS
(geostationary mobile
satellite standard)

circuit-switched cellular phone connections and the short message service (SMS).

A satellite air interface standard developed from GSM and formed by Ericsson, Lockheed Martin, UK Matra Marconi Space and satellite operators Asia Cellular Satellite and Euro-African Satellite Telecommunications.

GPS
(global positioning system)

A 'constellation' of 24 well-spaced satellites that orbit the Earth and make it possible for people with ground receivers to pinpoint their geographic location. The location accuracy is anywhere from 100 to 10 m for most equipment. Accuracy can be pinpointed to within 1 m with special military-approved equipment. GPS equipment is widely used in science and has now become sufficiently low-cost so that almost anyone can own a GPS receiver.

GSM
(global system for
mobile communication)

A digital mobile telephone system that is widely used in Europe and other parts of the world. GSM uses a variation of time division multiple access (TDMA) and is the most widely used of the three digital wireless telephone technologies (TDMA, GSM and CDMA). GSM digitizes and compresses data, then sends it down a channel with two other streams of user data, each in its own time slot. It operates at either the 900 or 1800 MHz frequency band.

Gateway

A combination of a software program and piece of hardware that passes data between networks. In wireless networking, gateways can also serve as security and authentication devices, access points, and more.

Hertz (Hz)

The unit for expressing frequency (f), a measure of electromagnetic energy. One Hertz equals one cycle per second.

Hotspot	A place where users can access Wi-Fi service for free or a fee.
Hotzone	An area where users can access Wi-Fi service free or for a fee.
HSDPA (high-speed downlink packet access)	A technology that enables faster data transmission via W-CDMA networks. HSDPA is a 3GPP Release 5 Standards-compliant modification to the UMTS air interface.
Hub	A multiport device used to connect several PCs to a network.
iDEN (integrated digital enhanced network)	A wireless technology from Motorola combining the capabilities of a digital cellular telephone, two-way radio, alphanumeric pager and data/fax modem in a single network. iDEN operates in the 800 and 900TMHz and 1.5 GHz bands and is based on time division multiple access (TDMA) and GSM architecture.
IEEE	Institute of Electrical and Electronics Engineers (www.ieee.org), in New York, is an organization composed of engineers, scientists and students and is best known for developing standards for the computer and electronics industry. In particular, the IEEE 802.xx standards for local-area networks are widely followed.
IMT-2000 (International Mobile Telecommunications-2000)	The United Nations International Telecommunications Union (ITU) globally coordinated definition of 3G covering issues such as frequency spectrum use and technical standards.
Internet protocol (IP)	A set of rules used to send and receive messages at the Internet address level.
IP address	A 32-bit number that identifies each sender or receiver of information that is sent across the Internet. An IP address has two parts–an identifier of a particular network on the Internet and an identifier of the particular device (which can be a server or a workstation) within that network.

IPsec (Internet protocol security)	A framework for a set of protocols for security at the network or packet processing layer of network communication. Earlier security approaches have inserted security at the application layer of the communications model. IPsec is useful for implementing virtual private networks and for remote user access through dial-up connection to private networks.
ISM band (industrial, scientific and medical band)	The 2.4 GHz band available worldwide on a secondary basis as long as the spread spectrum modulation techniques are used at relatively modest power levels.
ITU (International Telecommunications Union)	An institute based in Geneva, which is a pioneer in creating worldwide agreements and consultations on international standards.
Java	A programming language designed for use in the distributed environment of the Internet. Designed to have the look and feel of the $C++$ language, it is simpler to use than $C++$ and enforces an object-oriented programming model. Java can be used to create complete applications that may run on a single computer or be distributed among servers and clients in a network.
LAN (local area network)	A high-speed network that connects a limited number of computers in a small area, generally a building or a couple of buildings.
MAC (media access control)	A unique identifier that can be used to provide security for wireless networks. All Wi-Fi and WiMAX devices have an individual MAC address hard-coded into them.
MAN (metropolitan area network)	A network that interconnects users with computer resources in a geographic area or region larger than that covered by even a large local area network (LAN) but smaller than the area covered by a wide area network (WAN).

Mbps
(millions of bits per second
or megabits per second)

A measure of bandwidth (the total information flow over a given time) on a telecommunications medium. Depending on the medium and the transmission method, bandwidth is also sometimes measured in kbps (thousands of bits or kilobits per second) range or the Gbps (billions of bits or gigabits per second) range.

NFC
(near-field communication)

A technology that enables short-range communication networks between consumer devices incorporating an NFC interface, and is set to greatly improve the way consumers access data and services wirelessly.

NIC
(network interface card)

A type of PC adapter card that works without wires (Wi-Fi) or attaches to a network cable to provide two-way communication between the computer and network devices such as a hub or switch.

P2P (peer-to-peer) network

Also known as *ad-hoc* mode, this is a network of computers that has no server or central hub. Each computer acts both as a client and network server. It can be either wireless or wired.

PAN (personal area network)

A casual, close-proximity network where connections are made on the fly and temporarily. Meeting attendees, for example, can connect their Bluetooth-enabled notebook computers to share data across a conference-room table, but they break the connection once the meeting is over.

PCMCIA
(personal computer memory
card international association)

An industry group organized in 1989 to promote standards for a credit card-size memory or I/O device that would fit into a personal computer, usually a notebook or laptop computer. The PCMCIA 2.1 Standard was published in 1993. As a result, PC users can be assured of standard attachments for any peripheral device that follows the standard. The

PCS
(personal communications
services)

initial standard and its subsequent
releases describe a standard product, the
PC card.

A wireless phone service similar to cellular telephone service but emphasizing
personal service and extended mobility.
It is sometimes referred to as digital
cellular (although cellular systems can
also be digital). As a mobile user moves
around, the user's phone signal is picked
up by the nearest antenna and forwarded
to a base station that connects to the
wired network.

PDA
(personal digital assistant)

A term for any small mobile handheld
device that provides computing and
information storage and retrieval capabilities for personal or business use,
often for keeping schedule calendars
and address book information. The
term 'handheld' is a synonym. Many
people use the name of one of the popular PDA products as a generic term.
These include Hewlett-Packard's Palm-
top and 3Com's PalmPilot.

PDC
(personal digital cellular)

A TDMA-based Japanese standard
operating in the 800 and 1500 MHz
bands

PHS (personal handy system)

A TDD TDMA Japanese-centric system
that offers high speed data services and
superb voice clarity. It is really a WLL
system with only 300 m to 3 km coverage.

POTS
(plain old telephone system)

This provides a standard analogue telephone service.

PSK (phase-shift keying)

A method of transmitting and receiving
digital signals in which the phase of a
transmitted signal is varied to convey
information.

Plug and Play

A feature of a computer system which
enables automatic configuration of add-
ons and peripheral devices like wireless
PC cards, printers, scanners and multimedia devices.

RF (radio frequency) — Frequency within the electromagnetic spectrum associated with radio-wave propagation.

Range — The distance a wireless signal can reach.

Repeater — A device that receives a radio signal, amplifies it and retransmits it in a new direction. Repeaters are used in wireless networks to extend the range of base-station signals, thereby expanding coverage, within limits, more economically than by building additional base stations.

Roaming — The ability to move from one access point coverage area to another without losing connectivity.

SDMA (space division multiple access) — Thought of as a component of 3G digital cellular/UMTS

SDR (software defined radio) — Wireless communication in which the transmitter modulation is generated or defined by a computer, and the receiver uses a computer to recover the signal intelligence.

SIM (subscriber identity module) — A smart card inserted into GSM phones that contains telephone account information. It lets you use a borrowed or rented GSM phone as if it were your own. SIM cards can also be programmed to display custom menus on the phone's readout.

SoHo — A term for the small office or home office environment and business culture. A number of organizations, businesses and publications now exist to support people who work or have businesses in this environment. The term 'virtual office' is sometimes used as a synonym.

Satellite broadband — A wireless high-speed Internet connection provided by satellites. Some satellite broadband connections are two-way–up and down. Others are one-way, with the satellite providing a high-speed downlink and then using a dial-up telephone connection or other land-based system for the uplink to the Internet.

Server	A computer that lets other computers and devices on a network share its resources, including print servers, Internet servers and data servers. A server can also be combined with a hub or router.
Spectrum allocation	The range of frequencies designated by a National Telecommunications Regulatory Authority for a category of use or uses.
Switch	A network device that selects the path that a data packet will take to its next destination, ensuring optimal network performance. The switch opens and closes the electrical circuit to determine whether and to where data will flow.
Tablet PC	A personal computer that allows a user to take notes using natural handwriting on a stylus- or digital pen-sensitive touch screen instead of requiring the use of a keyboard. The tablet PC is similar in size and thickness to a yellow paper notepad.
TDMA (time division multiple access)	A technology used in digital cellular telephone communication that divides each cellular channel into three time slots in order to increase the amount of data that can be carried.
Triple DES	An encryption method which, like DES, operates on 64-bit data blocks. There are several forms, each of which uses the DES cipher three times. Some forms use two 56-bit keys, some use three.
UMTS (universal mobile telecommunications service)	A 3G broadband, packet-based transmission of text, digitized voice, video and multimedia at data rates up to 2 Mbps. It will offer a consistent set of services to mobile computer and phone users no matter where they are located in the world. Based on the global system for mobile (global system for mobile communication) communication standard, UMTS, endorsed by major standards bodies and manufacturers is the planned standard for mobile users.

VoIP (voice-over IP)

A technology that supports voice transmission via IP-based LANs, WANs and the Internet.

VPN (virtual private network)

A way to use a public telecommunication infrastructure, such as the Internet, to provide remote offices or individual users with secure access to their organization's network. A virtual private network can be contrasted with an expensive system of owned or leased lines that can only be used by one organization. A VPN maintains privacy through security procedures and tunneling protocols and encryption.

VSAT
(very small aperture terminal)

An earthbound station used in communications of data, voice and video signals, excluding broadcast television, consisting of two parts: a transceiver placed outdoors in direct line-of-sight to the satellite, and a device placed indoors to interface the transceiver with the end user's communications device, such as a PC.

WAN (wide area network)

A geographically dispersed telecommunications network. The term distinguishes a broader telecommunication structure from a local area network or LAN. A wide area network may be privately owned or rented, but the term usually connotes the inclusion of public (shared user) networks.

WAP
(wireless application protocol)

A specification for a set of communication protocols to standardize the way that wireless devices, such as cellular telephones and radio transceivers, can be used for Internet access, including e-mail, the World Wide Web, newsgroups, and Internet relay chat (IRC). While Internet access has been possible in the past, different manufacturers have used different technologies. In the future, devices and service systems that use WAP will be able to interoperate.

WCDMA (wideband code-division multiple access)	An ITU standard derived from code-division multiple access (CDMA), officially known as IMT-2000 direct spread. WCDMA is a 3G mobile wireless technology offering much higher data speeds to mobile and portable wireless devices than commonly offered in today's market.
WEP (wired equivalent privacy)	A security protocol, specified in the IEEE wireless fidelity (Wi-Fi) standard, 802.11b, that is designed to provide a wireless local area network (WLAN) with a level of security and privacy comparable to what is usually expected of a wired LAN.
Wi-Fi (short for 'wireless fidelity')	The popular term for a high-frequency wireless local area network (WLAN). The Wi-Fi technology is rapidly gaining acceptance in many companies as an alternative to a wired LAN. It can also be installed for a home network. Wi-Fi is specified in the 802.11b specification from the IEEE and is part of a the 802.11 series of wireless specifications. The 802.11b (Wi-Fi) technology operates in the 2.4 GHz range offering data speeds up to 11 Mbps.
WiMAX Forum	A coalition of wireless-industry leaders committed to the open interoperability of all products used for broadband wireless access based on 802.16 IEEE standards.
WISP (wireless Internet service provider)	An organization providing wireless access to the Internet.
WPA (Wi-Fi protected access)	A successor to WEP as a security protocol endorsed by the Wi-Fi Alliance. WPA is an interim security enhancement designed to be forward-compatible with IEEE 802.11i.
WLAN (wireless LAN)	A network in which a mobile user can connect to a local area network (LAN) through a wireless (radio) connection. A standard, IEEE 802.11, specifies the technologies for wireless LANs. The standard

	includes an encryption method, the wired equivalent privacy algorithm.
Wireless loop	A wireless system providing the 'last mile' connectivity; that is, the last wired connection between the telephone exchange and the subscriber's telephone set (which can be up to several miles in length). Traditionally, this has been provided by a copper-wire connection.
WRC (world radiocommunication conference)	An international conference organized by the ITU at which standards and interference issues are discussed at the intergovernmental level.
Yagi antenna	Also known as a Yagi-Uda array or simply a Yagi, a unidirectional antenna commonly used in communications when a frequency is above 10 MHz. This type of antenna is popular among amateur radio and citizens' band radio operators.

Bibliography

1. Helin, J. and MediaLab. 2003. Challenges for the Future Mobile Operator Consumer Business, mBusiness 2003 Conference; available at www.mBusiness2003.org
2. R4B, WiMAX Opportunity – Global, 2005, Resource4Business.
3. Jonason, A. 2002. Innovative pricing effects: theory and practice in mobile Internet networks. *European Journal of Innovation Management*, 5(4): 185–193.
4. Mylonopoulos, N.A., Sideris, I., Fouskas, K. and Pateli, A. 2002. Emerging Market Dynamics in the Mobile Service Industry White Paper. The Mobicom project.
5. Nordstram, B. 2001. Value, Content, Partnerships and Revenues in the Mobile-Internet Era. Advanced Internet Applications, Mobile Commerce.
6. Pareek, D., 2005. *WiMAX: Taking Wireless to the MAX*. CRC Press.
7. Strouse, K.G. 1999. *Marketing Telecommunication Service: New Approaches for a Changing Environment*. Artech House, London.
8. Landa, R.T. 1997. Policy reform in networks infrastructure: the case of Mexico. *Telecommunications Policy*, 21(8): 721–732.
9. Lapuerta, C. and Mosell, B. 1999. Network industries, third-party access and competition law in the European Union. *Northwest Journal of International Law and Business*, 19(3): 454–478.
10. Lapuerta, C. and Tye, W.B. 1999. Promoting effective competition through interconnection policy. *Telecommunications Policy*, 23(2): 129–146.
11. Venkatesh, R. and Mahajan, V. 1993. A probabilistic approach to pricing a bundle of services. *Journal of Marketing Research*, 30(4): 508–526.
12. Yadav, M.S. and Monroe, K.B. 1993. How buyers perceive savings in a bundle price: an examination of a bundle transaction value. *Journal of Marketing Research*, 30(4): 350–358.

13. Competition Commission. 2003. *Vodafone, O₂, Orange and T-Mobile.* Competition Commission: London; available at: www.competition-commission.org.uk/reports/475mobilephones.htm#full

14. Grier, J. 2003, 802.16: a Future Option for Wireless MANs, www.wifiplanet.com/tutorials/article.php/2236611.

15. Kargl, F. Lawrence, E. and Zarumba, G.V. Introduction to the Minitrack on Wireless Personal Area Networks (WPANs); http://csdl2.computer.org/comp/proceedings/hicss/2004/2056/09/205690305.pdf

16. Bluetooth, The official Bluetooth website; www.bluetooth.com.

17. IEEE. Get IEEE 802; http://standards.ieee.org/getieee802/

18. Wi-Fi Alliance. What is Wi-Fi?; www.wi-fi-org./OpenSection/index.asp

19. WiMAX Forum. Welcome to the WiMAX Forum; www.wimaxforum.org/home

20. ZigBee Alliance. Control that Simply Works; www.zigbee.org/en/index.asp (accessed 12 July 2005).

21. IEEE. IEEE Standard 802.16: a Technical Overview of the Wireless MAN Air Interface for Broadband Wireless Access; http://grouper.ieee.org/groups/802/16/docs/02/C80216-02_05.pdf (accessed 12 July 2005).

22. Anderson, P. and Tushman, M.L. (1990). Technical discontinuities and dominant designs: a cyclical model of technological change. *Administrative Science Quarterly*, 35: 604–633.

23. Christensen, C. 1997. *The Innovator's Dilemma.* Harvard Business School Press: Harvard, MA.

24. Christensen, C. 2002. The rules of innovation. *Technology Review*, June: 32–38.

25. Gabriel, C. 2003. WiMAX: the Critical Wireless Standard – 802.16 and other broadband wireless options. Blueprint WiFi Monthly Research Report. ARC Chart Ltd.

26. Porter, M. 1985. *Competitive Advantage.* New York Free Press: New York.

27. Utterback, J. and Abernathy, W. 1975. A dynamic model of process and product innovation. *The International Journal of Management Science*, 3(6): 639–656.

28. WiMAX Forum. 2004. Business Case Models for Fixed Broadband Wireless Access based on WiMAX Technology and the 802.16 Standard.

29. Basso, M. 2003. Key Issues set the Agenda in Wireless and Mobile, Gartner, BLE-19-1628, 16 January 2003.

30. Deighton, N. 2003. Mobility Requires More Than Wireless-Enabled Devices, Gartner, AV-19-0285, 16 January 2003.

31. Dittner, P., Clark, B. and Deighton, N. 2003. Mobile Infrastructure, Technologies and Markets, Gartner, K19-0206, 13 January, 2003.

32. Troni, F. and Cozza, R. 2003. Personal Digital Assistants: overview, Gartner, DPRO-90774, 16 January 2003.

33. Troni, F. 2002. Smart Phones: a perspective, Garner, DPRO-99896, 2 December 2002.

34. Chavez, A. and Maes, P. 1996. Kasbah: an agent marketplace for buying and selling goods. In *Proceedings of the First International Conference on the Practical Application of Intelligent Agents and Multi-Agent Technology*, London, April 1996. Available at: http://citeseer.nj.nec.com/chavez96-kasbah.html

35. Bartlett, J.F. 2000. Rock'n' scroll is here to stay. *IEEE Computer Graphics and Applications*, **20**(3): 40–45.

36. Moukas, A. and Maes, P. 1998. Trafficopter: a distributed collection system for traffic information. In *Proceedings of Cooperative Information Agents '98*, Vol. 1435 of Lecture Notes in Artificial Intelligence. Springer-Verlag: Berlin Available at: http://lcs.www.media.mit.edu/projects/trafficopter/.

37. NTT DoCoMo; www.nttdocomo.com/home.html

38. Drucker, P. 1994. Knowledge Work and Knowledge Society: The Social Transformations of this Century, 1994 Edwin L. Godkin Lecture, 4 May Harvard University. Available from: www.ksg.harvard.edu

39. World Bank; www1.worldbank.org

40. Resource4business; www.r4b.in

41. WiMAX.com; www.wimax.com

42. IEEE website; www.ieee.org

Index

The Business of WiMAX   Deepak Pareek
© 2006 John Wiley & Sons, Ltd